Geography and Vision

Vision and visual imagery have always played a central role in geographical under-standing, and geographical description has traditionally sought to present its audience with rich and compelling visual images, be they the elaborate cosmo-graphic images of seventeenth century Europe or the computer and satellite imagery of modern geographical information science. Yet the significance of images goes well beyond the mere transcription of spatial and environmental facts and today there is a marked unease among some geographers about their discipline's association with the pictorial. The expressive authority of visual images has been subverted, shifting attention from the integrity of the image itself towards the expression of truths that lie elsewhere than the surface.

In *Geography and Vision* leading geographer Denis Cosgrove provides a series of personal reflections on the complex connections between seeing, imagining and representing the world geographically. In a series of eloquent and original essays he draws upon pictorial images – including maps, sketches, cartoons, paintings, and photographs – to explore and elaborate upon the many and varied ways in which the vast and varied earth, and at times the heavens beyond, have been both imagined and represented as a place of human habitation.

Ranging historically from the sixteenth century to the present day, the essays include reflections upon geographical discovery and Renaissance landscape; urban cartography and utopian visions; ideas of landscape and the shaping of America; widerness and masculinity; conceptions of the Pacific; and the imaginative grip of the Equator. Extensively illustrated, this engaging work reveals the richness and complexity of the geographical imagination as expressed over the past five centuries. It will appeal to all scholars with an interest in geography, history, art, landscape, culture and environment.

Denis Cosgrove is Alexander von Humboldt Professor of Geography at the University of California Los Angeles. A founding editor of the Journal *Ecumene* (now *Cultural Geographies*), his previous books include *The Palladian Landscape* (1993), *Social Formations and Symbolic Landscape* (2nd edn 1998), *Mappings* (editor, 1999) and *Apollo's Eye* (2001), which received the Association of American Publishers Professional and Scholarly Publishing Award in Geography and Earth Sciences. He is co-editor, with Veronica della Dora, of *High Places: Cultural Geographies of Mountains and Ice* (I.B.Tauris, forthcoming).

Geography and Vision
Seeing, Imagining and Representing the World

Denis Cosgrove

I.B. TAURIS

LONDON · NEW YORK

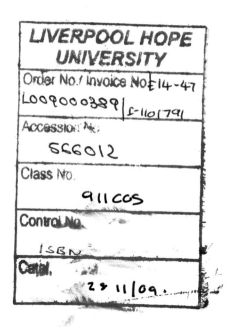
Published in 2008 by I.B.Tauris & Co Ltd

6 Salem Road, London W2 4BU
175 Fifth Avenue, New York NY 10010
www.ibtauris.com

In the United States of America and Canada distributed by Palgrave Macmillan
a division of St. Martin's Press
175 Fifth Avenue, New York NY 10010

International Library of Human Geography: Volume 12

ISBN: 978 1 85043 846 5 (HB)
ISBN: 978 1 85043 847 2 (PB)

A full CIP record for this book is available from the British Library
A full CIP record is available from the Library of Congress

Library of Congress Catalog Card Number: available

Typeset in Baskerville by Swales & Willis Ltd, Exeter, Devon
Printed and bound in Great Britain by TJ International Ltd, Padstow, Cornwall

Contents

CONTENTS

List of Illustrations

Plates (between pages 148 and 149)

Acknowledgements

This book has a complex genealogy. Its 13 chapters began life under diverse circumstances and at different times. All except the introductory essay and Chapter 12 originated independently from the present collection. I began to think of them as parts of a single project as I anticipated a period of sabbatical leave from UCLA in early 2006, and they appear here as revised and reworked during that period. Five have appeared in print previously, although in a different form from here, and in locations unlikely to be familiar to most readers of this collection.

Chapter 1 originated as my inaugural lecture delivered at Royal Holloway, University of London and published under the title *Geography and Vision* by the college in 1996. Chapter 2 began life as the Alexander von Humboldt annual lecture in the Geography Department, UCLA in March 2000. Chapter 3 originated as a lecture delivered at the 1997 Dumbarton Oaks Annual Symposium in Garden History. Chapter 4 is based on a public lecture delivered in April 2001 at the National Gallery of Canada, Ottawa, in connection with the exhibition 'Elusive Paradise'. Chapter 5 is a revised version of my chapter in James Corner and Alex MacLean's *Taking Measures: Across the American Landscape* (New Haven, Yale University Press, 1996, pp.3–13). Chapter 6 is a revision of 'Habitable earth: wilderness, empire and race in America' in David Rothenberg (ed.) *Wild Ideas* (Minneapolis, University of Minnesota Press, 1995, pp.27–41). Chapter 7 originated as a public lecture given in connection with the exhibition 'Ruskin and the Geographical Imagination,' curated by Colin Harrison and me at the Ashmolean Museum, Oxford in November 2000. Chapter 8 is based on a lecture delivered at the conference 'Victorian Europeans' at Royal Holloway, University of London in June 2005. Chapter 9 is a revised version of the chapter 'Historical perspectives on representing and transferring spatial knowledge' in M. Silver and Diana Balmori (eds) *Mapping in an Age of Digital Media* (Wiley-Academy, 2003, pp.128–37). Chapter 10 is a revision of a chapter with the same title in Janet Abrams and Peter Hall (eds) *Else/where: Mapping: New Cartographies of Networks and Territories* (Minneapolis, University of Minnesota Press, 2006). Chapter 11 was originally delivered as a paper to the conference 'From the South Sea to the Pacific Ocean: Conceptualizing the Pacific 1500–1945' held at the Huntington Library, San Marino, California in March 2003. I thank all those involved in these various projects and events for the initial opportunity they provided for me to formulate and develop the ideas that have formed this book.

I should like to thank Veronica della Dora for her research assistance, editorial work and invaluable academic advice in the preparation of the book. It would not have appeared without her support and advice. Chase Longford ensured the quality of the illustrations with his accustomed skill and good humour. David Stonestreet was a loyal and supportive editor. I should also like to thank my colleagues in the UCLA Geography Department for their practical and emotional support during the difficult times that accompanied the final stages of preparing this manuscript.

Introduction: Landscape, map and vision

Landscape and *map* are geographical keywords. They act as conceptual pillars for this collection of essays. For some, landscape is the principal object of geographical science, to which the great twentieth-century American geographer Carl Sauer and many of his students devoted lifetimes of field study and teaching.[1] Since Sauer's death in 1975 a generation of geographers has rediscovered landscape, conceptually repolishing it and extending its reach well beyond the localized material expressions of land and life that had formerly captured geographical interest. Attention to landscape is now widely shared across a range of disciplines, for example in art history and architecture, literature and history.[2] The map, too, is a fundamental and conventional attribute of geography, denoting the person and work of the geographer in artistic iconography from at least the time of Vermeer.[3] But the map as a tool of geographical science, and mapping as an active practice in the complex construction and communication of spatial knowledge, have also been subjected to concentrated reflection, criticism and debate over the past quarter century, so that, like landscape, the map and mapping are now the subject of wide, interdisciplinary interest and use.[4]

Both landscape and map are strongly pictorial terms, and this connects them intimately with the principal theoretical focus of this book: vision. Writers on landscape have consistently referred to the tension between the word's Germanic roots as a quasi-legal definition of an area characterized by the shared customs of an agrarian community, and its more modern and popular sense of a view of physical scenery, whose unity is aesthetic: a product of such visual qualities as composition, form, and colour.[5] German geographers and their Anglophone followers in the first half of the twentieth century forged these somewhat distinct usages into a geographical concept of landscape as the visible expression of rooted community, ecologically bound to the soil, and used it in defence of traditions of land and life felt to be threatened by modern society, economy, production methods

1

and communications.[6] In interwar Germany itself this intellectual project, like so many others, was subverted and distorted by the virus of National Socialism, souring the term *Landschaft* in post-war German geography. But concern about the loss of uniqueness and diversity among rural communities and the visible scenery that their labour has produced remains widespread, especially in those parts of Europe faced with a steeply declining agricultural sector and popular demand to preserve traditional rural scenery. These sentiments are strong enough to keep landscape high on both academic and political agendas.[7] While contemporary landscape studies in geography are acutely sensitized to the complex and always fluid social, economic, cultural and political currents that work alongside ecological and environmental processes to shape the visible scene, it is precisely the form, stability and appearance of the land to the eye – of the nationalist, the tourist, the local activist – that act as the spur to both the academic interest and political action surrounding landscape. The connection between landscape and vision remains powerful.[8]

Ties to the pictorial also remain strong in the case of maps and mapping, despite a parallel reorientation of their theoretical scope and meaning towards questions of process rather than form. Mid-twentieth-century cartography experienced its own drive to confine mapping practice and the map itself within the narrow strictures of instrumental science. A subsequent reaction has stretched the definition of a map and the practices of mapping well beyond the conventional sense of a scaled representation of measurable geographical facts located in absolute, Euclidean space, a subject I pursue in more detail in Chapters 9 and 10.[9] The distance between conventional usage and the metaphorical meanings of mapping as a cognitive spatial practice has shortened considerably in recent scholarship, so that all sorts of purely mental and imaginative constructs are now treated as maps, while supposedly objective and scrupulously accurate scale renderings of real-world distributions are regarded as inescapably dyed with ideological, psychological and other subjective hues.[10] Yet what remains consistent throughout the changing study of the map, within and outside geography, is its graphic reference. Mapping remains a way of representing the world; the map remains a visible image of the (or at least *a*) world.

The pictorial and geography

This collection of essays draws freely on the conceptual complexities of landscape and map, while hewing closely to their connections with the pictorial image and with vision more generally. In so doing it inevitably reaches beyond any narrow disciplinary definition of geography, benefiting from and perhaps contributing to the contemporary interchange of disciplinary perspectives that has so enriched our understanding of environment, culture and meaning. In some chapters, landscape and map are treated together, and the focus of attention is in part on their interactions. As I outline below, these interactions have been historically strong. Most chapters, however, concentrate more exclusively on aspects of either landscape or mapping. But throughout, it is a shared pictorial reference and, by extension, a common association with the role of vision in geographical knowledge and practice that unifies this collection. My aim is to reveal and interrogate some of the myriad ways in which the vast and varied earth known to humans, in whole or in part, and at times in extension to spaces beyond the earth's surface, has been imagined and represented as a place of human habitation, care and desire, and to do so through the medium of graphic images. That these images are taken exclusively from the Western tradition and from the early modern and modern periods reflects the limitations of my own scholarly skills rather than any claims for exceptionalism or privilege on the part of 'the West'. The term 'graphic images' denotes an intentionally loose category that includes maps, sketches, paintings and photographs, all of which are used to illustrate the individual studies. It also encompasses some written descriptions, especially those such as John Ruskin's which pay special attention to the poetics of place and landscape form. What holds the category together is the capacity of such images to represent geographical vision in the dual sense of communicating eyewitness knowledge and interpretation of geographical realities, and of conveying the forms and ideas, the hopes and fears that constitute imagined geographies. In both these modes of geographical cognition, and in their constant interactions, graphic and pictorial images play active and creative roles that take the significance of representation well beyond mere transcription of spatial and environmental facts.

That the association between geography and maps remains strong in the popular mind, whether as rote learning of geographical facts by colouring or locating places on maps, or more recently as the manipulation of spatially referenced data in Geographical Information Science (GIS), is a sign of the continued significance of graphic images in shaping geographical knowledge. But among professional geographers, especially those concerned with social and cultural questions, there is today a marked unease about the association between geography, the pictorial and vision. This is expressed in various ways and with different degrees of justification. One expression results from the embrace of social theory, largely derived from continental European philosophers and students of what in France are called *les sciences humaines*, that has so enchanted Anglophone geographers in the decades since David Harvey instructed his colleagues to post on their office walls the phrase: 'by our theories you shall know us'. Theory, by definition, is conceptual: it works through sets of propositions and logical arguments, removed by one or more degrees of abstraction from the empirical materials and relations that it concerns. Its mode of inscription is formal language: either a refined and clarified version of common spoken and written language or, at its most sophisticated, syllogistic logic and mathematics. Pictorial images, which are more synthetic and expressive than analytic and logical, are not in themselves theoretical. That is not to claim that there are no theories of images, or that images are not informed by theories: both have been matters of intense critical reflection, not least within geography. But the critical stance that today frames cultural geography's relationship with pictorial images has itself tended to subvert their expressive authority, distracting attention from the integrity of the image itself towards the conditions of its production, circulation and reception, and potentially reducing it to a mere expression of truths that lie elsewhere than on its surface.[11] A further consequence of the embrace of theory has been the elevation of text over image, both as authoritative source and preferred mode of communication within contemporary geography, after a period in the earlier part of the twentieth century when the map, for example, was regarded as the origin of geographic questions and the most appropriate expression of their answers.[12] The disappearance of the map in much social and cultural geography and an often uncritical approach to the actual use of pictorial images as illustrations are indicative of this shift.

A specific expression of the strained relations between geography and the pictorial image has been the rise of 'non-representational theory'.[13] This term now incorporates a wide range of ideas and practices, unified by a recognition that human cognitive and affective ties to the world do not operate solely through the sense of sight and that geographical knowledge is variously performative more than simply representational. Sight is commonly figured by non-representational theorists as having overly dominated rational and scientific ways of understanding and reflecting on human relations with the physical and material world.[14] The distrust of vision is rooted in a number of theoretical dispositions. Second-wave feminism and postcolonial theory in the 1980s and 1990s argued that the dominant ways of seeing that had emerged with perspective theories in the West during the fifteenth century, and their relationship with representation in painting, photography and moving pictures (as well as with cartography), were phallocentric, colonialist and calculating. 'The gaze', as this mode of seeing and its related forms of representation were sometimes termed, is inescapably voyeuristic, domineering and exploitative; it demanded resistance.[15] Although often crudely stated (and the subject of considerable criticism within feminist theory itself), this argument opened visual studies to a wide set of ideas about the complexity and complicity of vision: the differences for example between looking, glancing, seeing and staring, and about the social and historical nature of sight. It also introduced psychoanalytic theories about relations between seeing, the reflected image, and identity. Partly through such insights and partly through scientific advances in neurology, this has shifted theoretical attention to relations between the sense of sight and other bodily senses such as hearing, touch and taste, acknowledging that the eye is always embedded in a fleshly body. Indeed, the conventional classification of five discrete senses, while intuitively appealing in that they relate to distinct organs and locations in the body, seems to be heading towards the scientific oblivion of the four Aristotelian elements, as we recognize that no one sense operates independently of the others and instead regard the body as a muscular-skeletal, organic and neurological whole.[16] The implications of these insights for geographers interested in spatial cognition and conduct are considerable, especially when connected to questions of personal and social identity that are no longer regarded as essential aspects of the human self, but as constructed and labile, made and altered in the performance of everyday life. In dealing with such questions, pictorial images can seem too static, mediated, distanced and restricted to have much bearing on everyday embodied activity.

These criticisms of the pictorial specifically and the visual more generally within geography are powerful and important. They relate more directly to the discipline's spatial than to its environmental tradition. To address them adequately is beyond the scope of this book. Indeed my concerns here are not principally theoretical at all, in the sense of the book being framed by a hypothesis about geography and vision that is worked through a set of arguments and examples. Rather, in the tradition of essay writing, which I discuss below, it is a collection of somewhat personal reflections on the complex connections between seeing, imagining and representing the world geographically. The reflections are not random: they are informed by a consistent set of interests and predispositions (rather than propositions) about geography and geographical knowledge. Central to these is an idea of vision.

Describing geography and vision

Vision is a complex word that incorporates both the ocular act of registering the external world, and a more abstract and imaginative sense of creating and projecting images. Neither of these meanings is simple, and we know that each has a social and historical as well as a purely physiological character. Taking the social and historical first, there are 'ways of seeing' that vary with individuals, genders, cultures and so on, and there are histories and historical geographies of seeing. Vision in the sense of active seeing is inescapable in the practice of geography. This statement is by no means as banal as it might appear. In the long and now largely superseded meaning of geography as the practice of exploring, reporting and recording the varied surface of the earth – its lands and seas, its climates and environments – eyewitness knowledge and verifying the truth of visual observation were crucial features of geographical science. For a period in the nineteenth century direct engagement in the work of exploration and eyewitnessing of 'other' places became a prerequisite for membership of the geographic fraternity,[17] and even when the geographer's role was restricted to compiling, evaluating and synthesizing the reports of others, verifying the truth of eyewitness accounts was a central geographical task.[18]

Geographical description, which performs the task of interrogating, synthesizing and representing the diversity of environments, places and peoples, has traditionally sought to present its audience with rich and compelling visual images. The map is one powerful way of achieving these goals. A conventional trope in the prefatory pages of the atlas for much of its history was that it offered the wonder of the vast and varied world in images that could be consulted in the privacy of one's study, negating the time, discomfort and potential danger of travelling to see such things in person. The same idea, differently expressed, was used to promote air photographs in the early years of flight, and it remains a part of the imaginative attraction of popular geographies such as *National Geographic Magazine* and Google Earth.

But the map is by no means the only medium through which geographical knowledge is conveyed. Written narrative and description hold as significant a place as cartographic representation in the history of geographical practice: the graphic can be textual as much as it can be pictorial. The two most significant sources of Ancient Greco-Roman geography, whose fifteenth-century translation into Latin so influenced the discipline's Renaissance revival at the very moment when the world known to Europeans was expanding to global dimensions, were Ptolemy and Strabo. Their works capture the two paths of geographical description.[19] The relatively short text of Ptolemy's *Geography* comprised a set of mathematical principles for mapping a spherical world in two dimensions, accurately locating places on the map, and relating maps of different scales one to another. The larger part of the work was made up of lists of coordinates from which maps of the known world could be produced. For Ptolemy, the descriptive product of geographical knowledge was the map. By contrast, Strabo's 'colossal work' that dates from about 150 years before Ptolemy's contained no maps.[20] His *Geography* was an encyclopaedic description of the various regions of the world known to Greeks and Romans and of the diverse landscapes and peoples therein. It sought to account for differences in social forms and cultural practices and did not shirk from offering explanations and making judgements about geographical difference, with a marked penchant for the curious, the marvellous and the exceptional.[21] If Ptolemy's sources were earlier mathematical geographers and scientists, navigators and travellers, Strabo's were as often poets and playwrights, most consistently Homer, whose writings offered a richly descriptive picture of peoples and places.

The distinction between cartographic and narrative forms of geographical description evidenced by Ptolemy and Strabo informed partially distinct streams in early modern geography. The 'discoveries' of the sixteenth and seventeenth centuries that gave geographical science such prominence in the early modern world were interpreted and disseminated cartographically by cosmographers such as Martin Waldseemüller, Giacomo Gastaldi and Gerard Mercator, and narratively by writers such as Sebastian Münster, Gianbattista Ramusio and the Hackluyts. This is not to suggest that written descriptions lacked illustrative maps or that maps were not often accompanied by text: Abraham Ortelius's *Theatrum Orbis Terrarum* (1570), often proclaimed as the first modern atlas and mentioned in a number of my chapters, comprises pages of maps on the recto and written descriptions of the mapped area on the verso. The great seventeenth-century cosmographic wall maps produced in Holland and France often incorporated extensive blocks of descriptive text, as well as mathematical diagrams and pictorial descriptions of seasons and elements, customs and costumes, cities and landscapes. These were truly baroque geographies: geographical *Gesamskunstwerken* (total artworks) from an age obsessed with wonders and with the mysteries and possibilities of light and vision.

But the relationship between map and text is rarely balanced, and over time, especially since the eighteenth-century adoption of 'plain style' in cartography that removed pictorial, textual and other 'decorative' devices from the map, the relationship between the different modes of geographical description has become increasingly distant.

The evolving distinction between the map as a scientific instrument and the text as the appropriate medium for geographic description has borne significantly on the role and relations of landscape in geography. One of Ptolemy's contributions to early modern geography was the distinction between geography as a science of the whole earth and its major divisions, in which mathematical consistency of scale and proportion was vital, and chorography as a mode of description in which truth to the individuality, personality and uniqueness of a place or region was the goal, and in which accurate scaling in relation to the larger spatial frame was not significant. Chorography was an art as much as a science. Renaissance chorographers combined mapping, landscape art and literary description (for example prospect poetry) in their compositions, initiating a tradition of geographical description that lasted well into the eighteenth century. Chorography would echo through the picturesque tradition of landscape, which in turn prompted landscape criticism, conservation, planning and design in the later nineteenth and twentieth centuries. The chapters here on John Ruskin and early university geography in Britain explore this theme in more detail. For now it is sufficient to underline the roles of mapping and landscape as complex media that bring together many forms of graphic description and represent particular forms of geographical vision.

But vision is more than the ability to see and the bodily sense of sight. Vision's meaning incorporates imagination: the ability to create images in the mind's eye, which exceed in various ways those registered on the retina of the physical eye by light from the external world. Vision has a creative capacity that can transcend both space and time: it can denote foreseeing as well as seeing. *Geography and Vision* makes explicit use of vision's multiple and overlapping meanings, drawing on the relationship between the role of images — graphic, pictorial and textual — in geography that I have been considering so far, and imagination: the human capacity to form mental images, especially of things not directly witnessed or experienced. Like any systematic body of knowledge, geography can never be confined to a wholly inductive rendering of empirical facts. The very acts of selection, classification and composition must be informed by deduction in the broadest sense even if the predicates and hypotheses — the imaginative work — informing deduction are unconscious and implicit. Much attention has been paid in recent years to geographical imagination and to imaginative geographies.[22] These terms are not easily or precisely defined but they register a recognition of the role played by images and imagination in shaping the ways that geographical information and understanding are constituted and circulated and in their material effects. It was the cartographer and historian of medieval geographical 'lore', John K. Wright, who alerted his professional colleagues to the importance of what he termed 'geosophy': the geographical ideas and beliefs, true and false, of every and any group of people, in short: geographical vision.[23] His paper appeared in 1947 at a time when most of his American geographical colleagues were committing themselves to a positivist ideal of objective science rather than to the subtleties of epistemology. Only after a complicated route through behaviourism, phenomenology and ideological critique, and partly through the promptings of colleagues in literature and the humanities, did geographers begin to think seriously about the place and significance of imagination in their work.[24] In recent years writing on geography and imagination has exploded. Attention has focused on such issues as the 'epistemological violence' done by colonial and post-colonial images of colonized places and peoples by their imperial oppressors, on the role played by geopolitical images and stereotypes of other peoples in framing political action, and on the complex ways that memory and desire operate collectively to shape the form and experience of urban places, tourist destinations, and revered or feared landscapes and spaces.[25]

Many of these themes can be detected running through the essays collected here, although none is dominant. This is in some measure because the questions of power and justice that inform so much current work in cultural geography and which press the study of imagination towards questions of social instrumentality are not central to my task here, in part because they elevate critique over the study of vision and images themselves. This is not to diminish the importance of critique, but rather to recognize that the play of imagination is a complex affair, which is as often concerned to reach for the good and to realize hopes and dreams (even when these are achieved at the cost of unacknowledged harm to others) as it is to consciously inflict harm on the world.[26] Thus these essays explore ways that geographical visions of landscape, place and peoples have mapped, measured and promoted ideas of the good, the true and the beautiful, without averting the eye from unnoticed aspects and unintended consequences that might be quite the opposite. They celebrate the richness and complexity of the geographical imagination as expressed by different individuals and groups at moments during the past five centuries, especially as that expression has taken expressive form in pictures and poetry.

Form and structure

I noted earlier that the book is composed of essays rather than a continuous narrative or theoretical development. The pieces have different origins. Some began as spoken lectures and addresses to varied audiences, others as written work. They date from more than a decade of teaching, speaking and writing. Each has been altered to meet the essay form I have adopted in this collection. The essay form suits my present goal in various ways. It is a style whose modern form originated with late Renaissance writers such as Michel de Montaigne and Francis Bacon, who themselves composed essays on geographical topics. Not only is the essay the most familiar and comfortable length and form for someone educated in the British university tradition, which until recently used it rather than the research paper as the principal method of scholarly training, but the essay allows a degree of freedom to the academic project itself. The essay does not rest its claims to originality on newly discovered or newly analysed facts or evidence; it does not claim to build upon previous research and thus demand exhaustive reference to a prior 'literature', nor does it set out to demonstrate or demolish a previously stated theory or hypothesis. Rather it addresses a question, a curiosity, an event or occurrence, and within a limited space of writing, seeks to elaborate the issue, expose its various facets and explicate its implications, all from the perspective of experience, reflection and prior study by the essay's author. If this appears loose and undisciplined by comparison with the research paper, monograph or experimental report, there is some truth in the observation. But the essay is subject to its own disciplines: of honesty to the facts it proclaims and evidence it adduces, of logic and integrity in argument, of length, style and coherent expression. Such discipline gives the essay an explicit rhetorical quality that is appropriate to my interest here in the role of images and poetics in geography.

The essays are grouped into sets of two by shared themes, in turn organized according to larger themes and a loose chronology. The opening essay which provides the title for the whole collection is a substantive, if broadly sketched, historical consideration of the role of images and imagination within geography that I have glossed theoretically in this introduction. Much of its subject matter is developed more fully in the essays that follow. Its opening survey and discussion sketch a broad historical and disciplinary frame for subsequent chapters. The second essay of the pair explores in more detail ideas of cosmography sketched out in the first. Until the seventeenth century, cosmography encompassed geographical study. A geocentric world, understood through Aristotle's physics, could not be described coherently outside the frame of the encircling heavens whose form and motion explained so much of terrestrial phenomena. This belief had considerable implications for mapping and picturing spaces, extending from the depths of earth to the furthest heavens. These implications reached well beyond science into the fraught political realms of faith and morals in the early modern world. The seventeenth-century century collapse of cosmography, brought about in part by geographical discovery and a changing image of the earth's surface, excluded from geography scientific consideration of the heavens. Modern geography inherits that exclusion, but that is being put to the test as humans leave various cultural footprints across the inner planets of our solar system.

The succeeding four essays all deal with landscape. They focus for the most part on the early modern world, the first period of European oceanic navigation and colonization, when geography was the central science whose mode of operation – discovery, or bringing to the witness of the eye that which had previously been hidden – gave to Western science in general its principal trope. Rather than recording or assessing the processes, facts or consequences of discovery, however, my focus is on the role of vision and imagination in shaping and giving meaning to discovery. In this context it is important to remember that the rhetoric of 'new worlds' that informed so much of Renaissance geographical discovery was deeply rooted in a reassessment of 'old worlds', which for many humanists involved renewal, rebirth and renovation. This in turn required a critical exploration of the past, which meant principally a classical antiquity that was then known primarily through its artistic expressions in art and architecture, literature and poetry. 'Gardening the Renaissance world' and 'Mapping Arcadia' explore ways that geography and history within Europe themselves served as sources and vehicles of desires and fears, made urgent in the confrontation with unknown spaces and peoples across the Atlantic. The second pair of essays, 'Measures of America' and 'Wilderness, habitable earth and the nation' moves the locus of discussion across the Atlantic to examine ways that landscape ideas have played out in the early and continued shaping of American space and American nature, specifically in the United States.

The middle pair of essays brings us forward in time to the nineteenth century and concentrates on the context and writings of one individual, John Ruskin: landscape writer, art and social critic and moralist, whose work strongly shaped my own geographical imagination as a young academic. Ruskin's life spanned the years of Britain's radical geographical transformation into the first urbanized-industrial nation in history. Many of the issues concerning landscape, culture and modernity that are still being worked out globally today, as the passage to urban life is experienced by all the world's peoples, were the subject of Ruskin's thoughtful and passionate scrutiny, addressed through the close interrogation of images and communicated in an intensely poetic style. In 'The morphological eye' I examine the principles of Ruskin's landscape vision and show how much it had in common with the 'new' geography established in late nineteenth-century Oxford, under the leadership of Sir Halford Mackinder and with the active participation of Patrick Geddes' acolyte, Andrew John Herbertson. In 'Ruskin's European visions' I turn the attention to the critic's ideas of European geography and the various imaginative mappings he generated, largely out of a romantic historiography of art and architectural creativity. If Ruskin's cartography today seems quaint at best, and ethnocentrically conservative at worst, I try to show that the questions he addressed about people and territory in Europe have not disappeared; indeed in the opening decade of the twenty-first century they seem as urgent as ever.

Ruskin's use of maps introduces the second major theme of the book, the reciprocal relations between cartography and geographical vision. 'Moving maps' is a reflection on changes in the way we think of cartography as a practice and maps as instruments. After reviewing the application of critical theory to maps and mapping, it focuses on the growing appreciation of the active and creative role that they both play in shaping and representing geographical imagination. 'Carto-city' develops these ideas in the case of urban mapping, tracing the connections between the rhetoric of urban cartography and utopian visions, from the Renaissance to our own days of flexible digital mapping, and of site-specific art practices that make use of the power and creative potential of the maps and mapping.

The closing two essays depart from the narrow focus on landscape and mapping, in order to concentrate on 'metageographical' features at the global scale. 'Seeing the Pacific' traces various ways that an oceanic space peppered with islands and covering nearly half the globe has come to be conceived as a single geographic region. The attention is on the nineteenth and twentieth centuries and specifically on American conceptions of the Pacific as they related to the shaping of an American cultural and geo-political vision of the nation and its role on the world stage. Maps and cartoon images, especially popular images produced during the Second World War campaign against Japan in the Pacific, provide substantive evidence of the role pictorial images can play in shaping a widely held imaginative geography. The last chapter, 'Seeing the Equator', is written in a lighter vein. It essays a cultural geography of a globally significant place which lacks dimension, cannot be seen, and yet is the stuff of diverse geographical visions. Drawing on an eclectic range of scientific, literary and cartographic sources it seeks to express and communicate something of the imaginative hold that this uniquely geographical phenomenon still possesses in the minds of so many people. In a world too often regarded as disenchanted and despoiled, the Equator's enduring grip on our imagination is a cause of wonder and perhaps of hope. It is certainly a powerful reminder of the enduring significance of geographical images and geographical vision and thus a fitting subject with which to close this collection.

I Geographic and cosmological visions

1 Geography and vision

The heavens declare the glory of God, and the firmament sheweth his handywork.

(Psalms 19:1)

Geographical inscription is simultaneously material and imaginative, shaping landscapes out of the physical earth according to human intentions: both the demands of practical existence and visions of the good life. Geographical representations – in the form of maps, texts and pictorial images of various kinds – and the look of landscapes themselves are not merely traces or sources, of greater or lesser value for disinterested investigation by geographical science. They are active, constitutive elements in shaping social and spatial practices and the environments we occupy. Reading landscapes on the ground or through images and texts as testimony of human agency is an honourable contribution for cultural geography to make towards the humanities' goals of knowing the world and understanding ourselves: to the examined life.

The first verse of Psalm 19 introduces a tradition of linking vision – the cosmic *glory* of God – with order, his *handiwork* within geographical space. Insofar as human geography concerns the natural or physical world, geographical space remains space that can be seen, or at least *visualized*. The different resonances of *vision* and *visualizing* constitute my theme. At its simplest, geographical vision ranges over space, but geographical space exists in historical time, involving contingent relations between an active observer and the field of observation. Vision is more than optics and perception, as I have already pointed out, and cultural geographers have carefully dissected many of its ideological complexities and tensions.[1] Vision also implies reworking – and pre-working – experience in the world through imagination, and imagination's expression in the creation of images. Geography (geo-*graphia*) has always entailed making and interpreting images.

15

The scales of geographic vision

Psalm 19's hymn to the spaces of creation echoes historically, from St Augustine to John Ruskin, as a theological justification for scientific enquiry into external nature, and indeed for geographical exploration itself. As the science of 'discovery', geography provided an early template and a term for the central objective of the Baconian scientific project of bringing speculation and observation into a coherent discourse. The psalmist proclaims the world as a unitary creation, the work of a transcendent agent bringing form and order out of an original chaos of material elements. Simon Girault's 1592 book *Globe du monde* illustrates the psalm through the image of the spherical machine of the world turned through its diurnal revolution by an unseen hand. For cosmographers such as Girault, observation of God's creation reveals to human creatures the traces and the presence of divine handiwork. Through observation humans may learn to penetrate the always labile and contingent surface appearances of the world, and disclose its more enduring form and order. In the world's form and order, vision invokes evidence of design, a sacredness on earth.

This belief, that the world itself provides visual evidence of consistent order and design, whether sacred or secular, remains remarkably robust historically, in the face of insistent human experience of chance, contingency and unpredictability – deified as Fate or Fortune – as fundamental to the ways of the world.[2] Its articulation in the West owes as much to Greek as to Judaic thought. In his cosmological text, the *Timaeus*, Plato described an orderly creation whose hidden form may be rendered *visible* by mathematics, specifically geometry. Pythagoreans had advanced similar arguments, suggesting that geometrical 'harmony' was sensibly apparent to human eyes and ears (the German word *Stimmung* captures this sense better than any English term). Literally, geo-metry means 'earth measurement', a practice that the Ancient Greeks believed had originated in Egypt, where the annual Nile flood erased all visible property boundaries, necessitating astronomical observation – the sun's shadow cast across the earth by a gnomon (a stake or obelisk) – in order to fix and re-inscribe them upon the undifferentiated space left by the retreating waters. The historical truth of this account is of less importance than the fact that it binds the localized and domesticated human landscape of cultivated fields and farms to the divine firmament of rotating planetary bodies. Geometry maps celestial order onto terrestrial space and makes visible divine handiwork through an act of human imagination and intellection.

Conceiving the universe as a divine geometrical exercise focuses attention on its structure: on a skeletal system of points, lines and motions.[3] An alternative rendering, most obviously traceable to Aristotle's *Physics*, drew attention to the fabric of creation and to the earth as a body within it, of mass and volume, composed of tangible and corruptible elements, which required a temporal process of separation from an originally undifferentiated chaos. Light had to be separated from darkness, water from earth, for any semblance of balance and harmony to be established and sustained within the corruptible elemental sphere. This cosmogony too could be observed along the Nile, as a cultural landscape of bounded and differentiated spaces was annually recovered from the confused chaos of earth and water left by flood, and the earth once more made productive. These accounts of order in the body of the world both concern geographical form: geometry describes its spatial structure, and physics its elemental nature. Until Newton's time, the science that synthesized these accounts into a reasoned description of the material universe was cosmography.

Girault's *Globe du monde* is a work of cosmography. Like the Renaissance natural philosophers whose work he synthesized, Girault drew fairly promiscuously upon Christian doctrine, classical authority and empirical observation to generate his vision of the world.[4] The most important source and authority for cosmographers at Girault's time of writing remained Claudius Ptolemy, the first century CE author of two great scientific summaries of ancient locational knowledge: the *Almagest*, which described the pattern and movement of the heavenly bodies, and the *Geography*, which described the pattern of climates, lands and seas on the terrestrial globe. From these works Renaissance cosmographers such as Regiomontanus (Johann Müller), Martin Waldseemüller, Oronce Fine, Peter Apian or Sebastian Münster developed an ordered classification representing the earthly globe and its place in the universe at diminishing scales.[5] They established a threefold spatial hierarchy made up of cosmography, geography and chorography (Fig. 1.1) Cosmography dealt with the whole system of the geocentric universe, treated either mathematically as a system of forms and motions, or descriptively in its vastly varied and differentiated contents. Geography described the large-scale pattern of climates, lands and seas on the surface of the globe, and could be logically connected to cosmography mathematically and thematically through measured scale. Chorography limned local regions, or landscapes. Each of these sciences was a '-graphy' (*-graphia*), not a '-logy' (*-logia*). In other words, each of these discourses was constructed primarily through images rather than words; each appealed to the logic and authority of the eye and the inscription of form as much as to scriptural authority or formal syllogism.[6] A graphic imperative informed the work of the cosmographer, whose primary appeal was to vision: to make visible, at each of these descending scales, the order and harmony, and the contents of creation. Theologically speaking, if the text of God's glory was the Bible, elucidated by hermeneutics and written commentary, the image of God's handiwork was the visible world, whose order was exposed in mathematical diagrams, globes, maps and paintings. If the scholastic

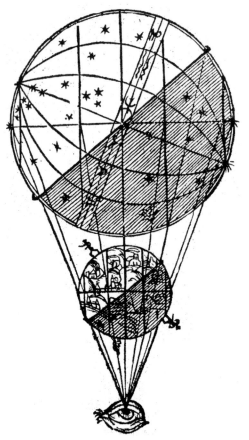

Figure 1.1 Peter Apian, Cosmographic relations of the spheres viewed from nowhere, *Cosmographicus Liber*, Landshut, 1524, fol. 1v (UCLA Library Special Collections)

might offer a commentary of the text of Genesis, the cosmographer's task was to reveal its cosmogonic narrative to the human eye, as is demonstrated by the opening images of Hartmann Schedel's *Nuremberg Chronicle* of 1493, building from the simplest geometrical figure of the circle to the great picture of God the Father enthroned at the centre of the ordered cosmos; or the mid-sixteenth-century Portuguese artist Francisco de Holanda's spectacular images of elemental parturition on the First Day, or the creation of sun and moon on the Fourth, all subject to the all-seeing eye of a Creator who is at once geometrical and corporeal (Fig. 1.2).[7]

I shall deal more fully with cosmography in Chapter 2. My concern here is those more terrestrial scales of geography and chorography to which

Figure 1.2 Francisco de Holanda, *Fiat Lux*: image of the first day of creation, *De Aetatibus Mundi Imagines*, 1543–73 (Biblioteca Nacional, Madrid)

geographical science is now perhaps too completely confined.[8] So exclusively has the contemporary secular vision become limited to 'spaceship earth', that many students of geography scarcely appreciate the significance of planetary arrangements for patterns of climate on the earth's surface. Yet until very recently, geographical learning began with the aid of an orrery or celestial diagram, intended to explain those purely conceptual lines of equator, tropics and polar circles described by the motion of the axially angled earth around the sun, whose geometry binds terrestrial and celestial spheres.[9]

Cosmographic vision

Cosmography intersects with geography in the great circles of equator, tropics and poles that encircle the globe and from which originate the ancient Greek idea of torrid, temperate and frigid zones, with all their ethnographic baggage. The desire to envision a consistent global order, beyond the scale of direct local experience, suggested not only a formal symmetry of global climate, but also of lands and seas (Fig. 1.3). Balancing continental landmasses in each hemisphere were believed to be surrounded by Ocean – a geographical descriptor that in Greek denoted a primeval slime of undifferentiated elements (*apeiron*) rather than open sea, a space wholly alien to the ordered landscapes of civilized urban life at the temperate centre of a habitable *oikoumene*.[10] The sixteenth-century French explorer Jacques Cartier and his settlers would pay the price of a Quebec winter for

Figure 1.3 Aurelius Theodosius Macrobius, Mappa mundi showing the climatic zones and balancing hemispheric continents separated by the Ocean Stream, In *Somnium Scipionis exposition*, Lyon, 1550 (UCLA Library Special Collections)

their inherited belief in parallel climatic zones, wherein a St Lawrence Valley January should resemble mild winter along the River Loire, because both occupy the same latitude. Similarly, until the eighteenth century, a great southern continent endured on world globes and maps, mute testimony to the lasting desire for symmetry between continental landmasses in the two hemispheres, while the original edition of Mercator's famous world map shows four rivers flowing from an open polar sea into a viscous ocean.

As geographical discovery revealed a pattern of lands and seas, environments and peoples quite different from that predicted by the theoretical geometries of astronomical geography, Europeans responded in two ways. Some quite ruthlessly imposed their visions of spatial order across conquered territories through the applied geometry of geodesy, survey and cartography, while others imagined ever more esoteric symmetries hidden beyond the earth's surface geography. I offer examples of each of these in turn.

Navigating seas, oceans and unmapped continental interiors depended upon astronomical observation and mathematical calculation, but they are practical activities than do not depend upon a geodesic framework. Throwing an imaginary lattice of longitude and latitude across the globe is, however, a prerequisite for projecting the spherical earth onto a flat map and accurately coordinating information about lands and seas into a single image. The grid's projective geometry made the earth visible and itself became a powerful stimulus to further visions of spatial order.[11] The grid's hubristic implications are suggested by the anthropomorphic wind heads, located at the cardinal points of many early world maps. Their interlocking gaze across the global surface signifies a divine, all-seeing, omnipresent eye, while their breath represents the spirit moving over waters and lands, animating the earth.[12] The eschatological significance of these figures is indicated by their regular conflation with angels, sounding the apocalyptic trumpet from the four corners of the earth. Yet the wind heads' gaze from the frame of the map is also a cartographic expression of the human conceit of special status in the order of creation: hovering between angelic and animal nature, able, in imagination at least, to escape the bonds of earth and encompass with the mind's eye its magnitude and plenitude. The grid itself resonates with the preternatural power these angelic wind heads signify.

The most consistent, extensive and permanent visible landscape trace of Europe's colonization of global landscapes during the half millennium between the mid-fifteenth and mid-twentieth centuries is the territorial inscription of cartography. The graticule, geodesy and the grid were practical tools of empire and modernity. Within a year of Columbus's first Atlantic voyage, Pope Alexander VI, claiming global authority from Christ's universal redemption, drew a meridian through a still virtual oceanic space to separate Spanish from Portuguese world empires, at that point no more

21

than flags and *padraõs* dotted along the map's uncertain coastlines. The line of Tordesillas, extended 35 years later by the Treaty of Zaragoza to encompass both hemispheres, was a geometrical *fiat* that still defines the linguistic geography of the South American continent.[13] The territorial divisions and agrarian landscapes of the United States and Canada, Australia and many parts of South America and Africa are a lattice of straight lines and right angles, testifying to the imposition of the cosmographic vision across widely varying environments and peoples. If the discovered globe did not fulfil its vision of geographical symmetry, Western authority would impose it through the agency of the surveyor and the cartographer.[14]

Less practically significant, but equally telling, is the continued desire for a natural symmetry within the earth's metageography.[15] By the early seventeenth century, when Francis Bacon was writing essays on so many of the themes that would dominate the modern world, including colonization and landscape and gardening,[16] voyages of discovery had shown decisively that the conjectural world map with its symmetrical pattern of northern and southern continents and oceans did not correspond in any obvious way to actual geography. My earlier reference to the Baconian project referred to his role as a metonym for exoteric, experimental science. Yet, like his contemporary Johannes Kepler, whose attempts to fit celestial motions into a harmonic geometry of nested polyhedrons was Pythagorean in inspiration, Francis Bacon in the *Novum Organum* also read intentional design into the forms made visible by the world map:

> similar instances are not to be neglected, in the greater portion of the world's conformation; such as Africa and the Peruvian continent, which reaches to the Straits of Magellan: both of which possess a similar isthmus and similar capes, a circumstance not to be attributed to mere accident.
>
> Again the new and the old world are both of them broad and expanded towards the north, and narrow and pointed toward the south.[17]

'A circumstance not to be attributed to mere accident', but rather to divine handiwork operating to sensible principles of symphonic and graphic order. In the years since Bacon wrote, a succession of thinkers has deployed ever more creative and erudite calculations to disclose symmetries within the geologically mobile planetary pattern of lands and seas inscribed on the geographic map. The geologist Charles Lyell developed a theory of antipodal asymmetry. The Russian earth scientist A. P. Karpinsky mapped geographical and geological homologies on a projection that pictured the earth's mountain chains forming a great tree branching across the globe; and in our own century, his compatriot Alexei Shulga attempted to use the eight faces of the cube to disclose congruence and symmetry in the pattern and topography of continents and oceans, suggesting that major relief forms such as mountain ranges and deep ocean trenches repeat themselves every 90 degrees along the equator. In the words of another Russian writing as recently as 1994:

> this trial of the earth's figure with symmetry testifies to the fact that not only the global relief of the earth is ruled by the symmetry and antisymmetry of the cube on the global scale, but that the internal organisation of the faces turns out to conform to the laws of periodicity and similarity with a spacing of 30 degrees.[18]

If such mystical musings seem more than a little dotty and ineffectual, the fascination with order in the global geographic pattern displays a tenacity that can be pernicious. Beyond their commercial value, the Suez and Panama canals were regarded by some nineteenth-century engineers (including Ferdinand de Lesseps himself) as contributions to a 'marriage' between the vital, progressive and masculine West and a passive, static and feminized East, while F. J. Turner's romantic historical vision of the Western frontier in nineteenth-century America was seized upon by German National Socialists, abetted by their geographers, convinced of Germany's destiny to expand into a symmetrical 'Wild East' in Poland and Byelorussia.[19] Within a more acceptable pantheon of modern geographers and explorers, we find Francis Younghusband literally spaced out in the Gobi Desert en route for Lhasa, or Franklin, Shackleton, Peary, Scott and Amundsen on sometimes suicidal treks designed to demonstrate elemental symmetries in oceanic space or to reach the poles – those global locations with no material existence, geometrical and geodesic points that even resist plotting with precision.[20] London's Royal Geographical Society still regarded Sir Ranulph Fiennes' symmetrical, bi-hemispherical polar hike of 1980 as significant enough to support financially and the British press reported it widely and in celebratory detail.

The fascination with planetary order and symmetry, whether exercised in the academic study with compass and ruler, or across polar wastes with dog sled and parka, offers evidence of the graphic imperative: the mesmerizing visual power of latitude and longitude lines or of continental and oceanic outlines inscribed on the global map. We know that in fact there are numerous poles (geographic, magnetic, geomagnetic), in distinct and mobile locations. We know too that the continent-ocean outline we map today is temporary, continuously altered in geological time by plate tectonics and sea level changes that we now know can be accelerated by our own actions. Geologically, continental shelves rather than coastlines offer more enduring patterns. The conjectural map of world geography in the immediate aftermath of the biblical flood drawn by the seventeenth-century Jesuit scholar Anasthasius Kircher seems almost to anticipate both the temporal instability and the geographical pattern of the modern continental map in its a speculative cartography of the post-diluvian global land area. His Jesuit vision seems to have been more comfortable than most in combining mobility and global symmetry: in the vault of the Society's principal church in Rome Kircher's contemporary, Andrea dal Pozzo, rendered the order's founder, Ignatius Loyola, at the centre of a vision of the four continents, being lifted through deep perspective space along rays of divine light to the risen Saviour.

Chorographic vision

The spatial vision of Counter-Reformation Rome extended logically from dal Pozzo's cosmographic panorama to the geographic and the local. The Vatican's great Gallery of Geographic Maps constitutes only one part of the cosmographer Egnazio Danti's scheme for the Belvedere and Terza Loggia, which included a tower of the winds with mechanical globes, a planisphere, Ptolemaic tables, and scientific instruments to calculate astronomical events for the longitude of Rome. It was an appropriately rhetorical project for Pope Gregory XIII, on whose revision of the Julian calendar Danti had advised, and whose resurgent Church was proclaiming universal mission in the worlds conquered by Spain.[21] The Gallery of Maps brings the cosmographic and geographical to the chorographic scale in a series of bird's eye views of Italian provinces.[22] In Ptolemaic language, such images are chorographies. In the *Geography*, Ptolemy had characterized chorography as the description of particular regions of the earth without concern for their precise relationship in scale or location to larger geographical patterns. Of primary importance in the chorography is rendering the character of a place. Ptolemy emphasized drawing and painting as skills in chorography, rather than mathematics and measurement, because its goal is to give a visual impression of the actual look of the land. Thus we see the Italian provinces in Danti's gallery from a bird's eye perspective that takes the eye from the axiometric to the planimetric in a single sweep over the image's surface. A similar and contemporary work is Cristoforo Sorte's 1556 map of the territory north-west of Venice, around the city of Treviso (*Plate 4*). The map illustrates lands suitable for improvement and reclamation, and inscribes a loose geometry of water channels and field boundaries across the landscape.[23] It is apparent from Danti's and Sorte's work that chorographic images follow a different logic of vision from that of the cosmographic diagrams and geographical maps. They are much closer to landscape paintings than to geographical maps, and indeed, a sharp distinction between chorographic mapping and landscape painting is historically impossible to make.[24] Yet there are compositional and technical features of chorographic maps that direct vision beyond the accidental and specific towards spatial harmonies and order, whether intrinsic or adventitious.

Each chorography is a bird's eye view of a small part of the earth's surface, cleverly combining different viewing positions. In addition to seeing the territory mapped out at a consistent scale, permitting accurate measurement of the distances between points, the observer gains a visual impression of distance and topography, as if looking though a picture or window frame at landscape scenery. As in painting, the frame itself serves as an ordering device, structuring and composing its contents.[25] Achieving the effect of distance depends upon sophisticated technical coordination of perspectival geometry. Perspective was of course the foundation of the Renaissance graphical revolution, and a technique rediscovered in the same locations and at the same time that Ptolemy's *Geography* was first translated into Latin and printed, and that the Portuguese began to navigate the open oceans.[26] Perspective seemed to many early modern thinkers the inherent constructional principle of space itself, proof that divine handiwork operated with geometrical consistency at all scales of creation, from the universe, through the earthly globe, to the immediately known local landscape. If the Roman architectural writer Vitruvius was to be believed, its geometry governed even the individual human body, whose elemental symmetries could be mapped according to the same mathematical logic that ordered the universe. This coherence is not so much demonstrated in words as envisioned through graphic images.[27] It is not surprising that when they came to transform landscapes – draining wetlands, channelling rivers or building roads, fortifications and cities – early modern engineers should seek to inscribe divine logic onto the terrestrial canvas. In the words of the sixteenth century architect Andrea Palladio:

> if we consider this beautiful machine of the world, with how many wonderful ornaments it is filled, and how the heavens, by their continual revolutions, change the seasons according as nature requires, and their motion preserves itself by the sweetest harmony of temperature, we cannot doubt, but that the little temples we make, ought to resemble this very great one, which, by his immense goodness, was perfectly completed with one word of His.[28]

So Palladio's villa designs are geometrical exercises, positioned to disclose the logic of the locally visible landscape in both their decoration and setting.

Yet perspective geometry is but one dimension of landscape representation. As in cosmography and geography, geometry provides only the skeletal structure of landscape, whose plenitude demands the more specific description seen in works that combine topographic with narrative specificity. Giovanni Bellini's painting of St Francis of Assisi's ecstatic vision on receiving the stigmata unites the fabric of the heavens, the physical earth and the fleshly human body through the rays of celestial light touching the saint's hands and feet (see Fig. 3.1). Bellini invites the viewer to share St Francis's vision of nature's variety and fecundity as well as its awesome power, in a perspective space that reaches from the stony wilderness of the hermit's cave, through a cultivated tapestry of fields and gardens, to the distant city. Bellini's landscapes are precisely located, chorographic landscapes. If St Francis prays at Monte Verna in Piemonte, elsewhere in Bellini's work Mary holds her infant, and cradles the body of her dead son, in the fields outside the city of Vicenza (with additional elements from Ravenna), set against the Venetian pre-Alps softened into blue by aerial perspective. Succeeding generations of painters, in Venice, the upper Danube cities and Flanders, developed and refined the genre of landscape painting to encompass both local topographies and vaster worlds into the restricted space of small, jewel-like images whose rich variety of detail earned them the label of cosmographies.[29] Indeed, painters of 'world landscape' such as Albrecht Altdorfer and Peter Bruegel the Elder worked closely with cartographers and shared with them the *furor geographicus* of sixteenth-century Nüremberg and Antwerp (see Plate 1).[30]

The detailed accuracy with which Bellini painted the individual plants and flowers surrounding St Francis was what John Ruskin, the nineteenth-century art critic, especially admired in early Venetian landscape painting. Speaking of a landscape by Giovanni Bellini's student, Titian, Ruskin remarks that:

> While every stamen of the rose is given, because this was necessary to mark the flower, and while the curves and large characters of the leaves are rendered with exquisite fidelity, there is no vestige of *particular* texture, of moss, bloom, moisture, or any other accident . . . nothing beyond the simple forms and hues of the flowers. . . . But the master does not aim at the *particular* colour of individual blossoms; he seizes the *type* of all, and gives it with the utmost purity and simplicity of which colour is capable.[31]

Ruskin's point is that the artist has a duty in painting landscape to reveal a beauty that is at once specific to, and inherent in, the forms of the natural world, a beauty that he regarded as immediate evidence of divine purpose. Like so many of his Victorian contemporaries who had been raised in the fervour of evangelical Protestantism, John Ruskin believed that divine handi-work was inscribed as precisely in the natural world as it was written in the Bible, and he constantly read the latter onto the former.[32] A bird's wing, a scree slope, a twig, a rock's fracture, or an Alpine peak all manifested similar curvilinear form(see Fig. 7.1). But it required concentrated vision to recognize the message of form, which explains the paradox with which Ruskin opens his instructional manual *The Elements of Drawing*:

the chief aim and bent of the following system is to obtain, first, a perfectly patient, and, to the utmost of the pupil's power, a delicate method of work, such as may ensure his seeing truly. For I am nearly convinced, that when once we see keenly enough, there is very little difficulty in drawing what we see; but, even supposing that this difficulty be still great, I believe that sight is a more important thing than the drawing; and I would rather teach drawing that my pupils may learn to love Nature, that teach the looking at Nature that they may learn to draw.[33]

The responsibility of the draughtsman and, above all, the artist, was to reveal graphically the form and order visible through all creation as the signature of divine handiwork.

More than once in his writings Ruskin turns directly to those lines from Psalm 19: 'the heavens declare the glory of God, and the firmament sheweth his handywork'. He is particularly gripped by the second clause. Firmament is an ill-defined word, he suggests, loosely meaning that part of the heavens which appears to have motion, in conventional cosmography the zone between the Moon and the fixed stars.[34] But Ruskin shifts its location to within the traditional elemental sphere, to refer to the zone of atmosphere immediately surrounding the earth, conventionally the lower region of air. Ruskin's source and inspiration lie in Genesis: 'the most magnificent ordinance of the clouds. . . . as a great plain of waters was formed on the face of the earth, so also a plain of waters should be stretched along the height of air, and the face of the cloud answer the face of the ocean'. It is at this mid-point, Ruskin believed, that a Christianized Apollo most directly and consistently traces His presence:

> by the firmament of clouds the golden pavement is spread for his chariot wheels at morning; by the firmament of clouds the temple is built for his presence to fill with light at noon; by the firmament of clouds the purple veil is closed at evening around the sanctuary of his rest.[35]

Given the significance of light as a unifying presence at each scale of the geographical vision, it is not surprising that Ruskin's despair at the moral and environmental corruption he believed he was witnessing in industrial England should be first visible in the firmament. In an 1884 lecture 'The storm cloud of the nineteenth century', Ruskin claimed to have identified over the course of the previous decade the increasing frequency of an atmospheric phenomenon which he refers to variously as a 'storm cloud' or 'plague wind', an undifferentiated gloom which could occur any time of year, even in high summer: a fitful, intermittent wind which shakes the leaves, a degraded cloud which blanches the sun 'like a bad half-crown in . . . a basin of soapy water'.[36] For Ruskin the explanation of this chill wind

27

and cloud was obvious: if human handiwork becomes degraded by an exploitative, immoral social and environmental order, then the signs of divine presence in nature will also be obscured from human vision.

Modernity and visions of geographic order

Ruskin's idea that the moral order in human society will be graphically reflected in the environmental order may today seem as quaintly old-fashioned as those earlier-mentioned ideas that continental symmetry, climatic patterns or planetary structures reveal divine handiwork. Yet, while expressed in less theological language, similar ideas endured through the twentieth century, and indeed played a central role in organizing and planning modern landscapes. In some respects, it was precisely against the moralistic – even sentimental – union of nature, art and vision, together with the historicist and decorative style of Victorian architecture associated with Ruskin, that the twentieth-century century Modern Movement so strongly reacted. Natural science rather than biblical doctrine was the authority to which the Modernist avant-garde in art, architecture, social improvement, philosophy and indeed geography appealed in the first two-thirds of the century. Among the guiding epistemological and aesthetic principles of the Modern Movement, broadly conceived, were order and planning. Picturing the land was a key element in the geographical expression of the rationalist Modern project.

In 1919 a new edition of the one-inch Ordnance Survey topographical map of England and Wales appeared. It was attractively coloured, folded, and had a distinctive cover showing a cyclist gazing over an iconic landscape of English downland and steeple-churched villages to the sea (see Fig. 6.2). For the first time since the national topographic survey was initiated more than a century earlier, the work was intended for public consumption rather than for strategic military use. Appropriately, it was called the *Popular* Edition. As Ellis Martin's cover illustrations suggested, it was intended to open English local landscapes to the eyes of the cyclist, the rambler and the motorist. In the 1920s and 1930s, *seeing* one's homeland became a national obsession in all European countries, promoted in Britain by such groups as the Ramblers' Association, the Cyclists' Touring Club, the Girl Guides and Boy Scouts; in Italy by the Touring Club Italiano and Giuseppe Bottai's Fascist Ministry of Education; and by various German youth associations that promoted *Wünderung*. Young people everywhere were encouraged to see and understand their local region, with the joint aims of maintaining the nation's natural and cultural heritage, developing nationalist pride, and improving their physical fitness.[37] All this was to be done rationally, through scientific planning. Nature Study and Geography were promoted as important parts of the school curriculum. Both emphasized field observation. Organized school trips to different parts of the country entered British high-school education for the first time, while map reading, field work, collecting and recording aspects of the local landscape with the aid of *Observer* books and I *Spy* manuals were all popular expressions of a widely shared belief that a national vision of moral and physical welfare derived from, and found expression in, landscape order. As I discuss in Chapter 6, university geographers in the early twentieth century grounded disciplinary practice in field observation and the development of the 'morphologic eye' that could discern structure and order in the visible patterns of landscape.[38]

One official expression of this orderly vision in England was the National Land Use Survey. The brainchild of London University geographer L. Dudley Stamp, the survey employed the schoolchildren of England and Wales to plot on six-inch Ordnance Survey maps the use of every parcel of land in the country. The survey lasted from 1930 until 1934 and resulted in a set of highly coloured one-inch scale summary map sheets, which became a basis for national land use planning in the aftermath of the Second World War.[39] The survey was to provide a tool for the vision of an *orderly* landscape that Stamp had inherited with so many of his colleagues from the influential and explicitly visionary social theorist, architect and planner, Patrick Geddes. Stamp's maps would make visible those places where the blight of Ruskin's storm cloud had visited England in the guise of mining, industry and unplanned urbanization, upsetting the supposed pre-industrial balance of town, village and country that so many mid-twentieth-century intellectuals regarded as the true expression of an English national genius in landscape. (see *Figs. 6.2* and *9.5*). An order supposedly evident in Georgian townscape and parkland, when God's handiwork in nature had been complemented by cultivation and building, was deemed to have been lost to the evils of urban industrialism in the succeeding century. This conservative narrative forms the structure of the most influential single twentieth-century text on landscape in England, W. G. Hoskins' *The Making of the English Landscape*.[40] Progressive modernists believed that electricity and scientifically advanced technologies offered an opportunity to recapture the 'natural' order without abandoning the benefits of human progress. Hydro-electricity particularly, which made use of the natural course of water along the topographic 'stream line', promised to fulfil this vision.[41] Fundamental to achieving it was to be the *plan*, and the plan had to be based on underlying form, to be revealed and exploited in the application of rational scientific principles.

A commonplace of Modernist planning in the post-war years that extended Patrick Geddes' vision of natural form as the foundation for social order was the notion that the great advances of twentieth-century atomic and sub-atomic physics had revealed non-Euclidean, time-space structures and movements as basic building blocks of nature, and that these structures and movements are repeated at every scale of creation. Developments in psychology suggested to some that they corresponded also to patterns and motions intuitively sensed by every human: 'form patterns in nature' as the mid-century landscape architect Gregory Keypes called them. Keypes suggested that 'we make a map of our experience patterns, an inner model of the outer world, and we use this to organize our lives. Our natural "environment" . . . becomes our human "landscape" – "a segment of nature fathomed by us and made home" '.[42] In similar vein but with a different reference point, Geddes' pupil, Patrick Abercrombie, creator of the influential post-war plan for London, appealed to Chinese geomancy or feng shui as a way of grasping intrinsic environmental and spatial order. New technology, particularly powered flight, which seemed to many mid-twentieth-century observers to offer a wholly new vision of spatial order, required a reworking of our 'graphic vocabulary':

> To convert this new environment into a human landscape, we need more than a rational grasp of nature. We need to map the world's new configurations with our senses, dispose our own activities and movements in conformity with its rhythms and discover in it potentialities for a richer, more orderly and secure human life.[43]

Modernist landscape architects sought to achieve exactly this. Geoffrey Jellicoe, for example, designed a series of paradigmatically Modernist spaces in Britain during the 1950s and 1960s. He used the 'organic' pattern of fields and footpaths revealed on an Ordnance Survey map of Gloucestershire as the geometrical motif for landscaping the reactors and storage areas of a nuclear power station at Oldbury on Severn, and sub-atomic structures as the basis for the campus of the newly designated York University and to figure Motopia: a cellular city of highways and high rise buildings intended to stretch over the wetlands and gravel pits of Staines to the west of London. In the 1962 design for the Rutherford High Energy Laboratory at Harwell in southern England, Jellicoe constructed great earthworks to contain the nuclear particle accelerator. These were modelled on the Neolithic barrows that break the visual line of the Berkshire Downs:

> In the subterranean laboratories at the foot of the hills the most advanced scientific studies as yet made by man are taking place. The scientist himself will tell you that the splitting of the atom leads to infinity, or as one scientist put it, 'to God'. The mathematical sciences have far outstripped the biological sciences, and this disequilibrium, as we all know, could lead to the eventual destruction of the human race. Opposed to this fearful intellectual development is the human body that is still the same as ever, and within this body, but very deep down

under layers of civilisation, are primitive instincts that have remained unchanged. It is probably true that the basic appeal of the rolling downlands is a biological association of ideas.[44]

Jellicoe's idea that splitting the atom leads to God is a Modernist restatement of the recurrent belief that the Creator's handiwork is made visible in the pattern and form of creation. Physics and mathematics, above all geometry, allow us to describe an ideal Platonic structure; the biological sciences reveal a more corporeal and flexible – Lucretian – order. The corollary for Jellicoe, as for Dudley Stamp, Patrick Abercrombie and other mid-century planners, was a moral imperative to realize the vision of order in the design of social space and landscape.

Jellicoe worried that understanding the physical and mathematical structures of nature had far outstripped the biological understanding of life, accounting in some measure for what he regarded as the soulless quality of Modernist planning's cold geometry. This is ironic, for the most remarkable shifts in the vision of created order in creation in the closing years of the last century came precisely from the biological sciences, especially from genetics. In many respects the view of the earth in the first decade of the third millennium is organic and biological rather than geometrical. Nowhere was this more dramatically expressed than in the familiar whole-earth photographs taken during the Apollo moon flights between 1968 and 1972 (Fig. 1.4) These offered a unique eyewitness image of the earth's surface, captured from a point high enough in the firmament to see the unshadowed globe.[45] They reduce the earth to the scale of a chorography, or landscape painting. What they reveal is not the geometrical order of the world map structured by lines of latitude and longitude, but the biological order of a planetary organism, a moist elemental sphere of earth, water and air, upon which the separation of elements and the bounding lines of geography appear smudged and unstable. Little wonder that such an image should stimulate visions of an animated biophysical totality, of *Gaia*.[46] In James Lovelock's influential late twentieth-century vision of a homeostatic earth, the graphic imperative reasserts itself. His thesis emerged from a project of speculating how one might characterize the nature of a life-bearing planet, but it gained much of its appeal from its association with the Apollo photographic images of Earth that possessed both the graphic structure and the pictorial plenitude of cosmography, even as they displaced conventional cosmography's anthropocentrism with a vision of self-sustaining and self-regulating planetary life. In achieving for the eyes of everyone the vision of earth, which was traditionally reserved uniquely for the Creator, global vision has been simultaneously secularized and re-enchanted.

Thinkers, artists and engineers since Plato have seen in spatial form and motion a cosmic order which they have sought to extend through the scales

Figure 1.4 AS17-22727 photograph of the earth from Apollo 17, 1972 (NASA, public domain)

of Creation to the natural world, to individual landscapes and to human bodies. Much of this discourse has found graphic expression, in the language of maps, diagrams and pictures. It is tempting to suggest that spatial harmonies resonate below the level of everyday consciousness and to build upon them visions of active intervention in the world. But I believe that this is to reverse the actual relationship between vision and action. Erwin Panofsky, the art historian who studied and reflected deeply on forms of graphic expression, spoke of a Western 'will to form', and we should hear in his phrasing the ominous echoes of Nietzchean will to power. We need to remain vigilant in the face of a constant temptation to leap from inscribing order and pattern into (geo)graphic images to inferring something more universal. Such inference can lead – indeed too often has led – inexorably to the temptations of applied geography: imposing a single vision across the wonderful variety and individuality of geographical actuality and human freedom. Yet the apparently insistent desire to disclose graphic order across contingent, haphazard, terrestrial spaces, and to envision in that order a transcendent, teleological artistry, reminds us that it is in imagination – in visions of the world – that science, faith and humanity coincide, as von Humboldt states so clearly in his introduction to the first volume of *Cosmos*. By illuminating earth's spatial and environmental dimensions, geographical visions can render their own service to the greater goal of the humanities: that of critically understanding ourselves and our participation in nature.

2 Extra-terrestrial geography

Alexander von Humboldt's best-known work is the five-volume *Cosmos: A Sketch of the Physical Description of the Universe*, published at the end of his life (part posthumously) between 1845 and 1862. Its first two volumes were rapidly translated into English, and the translation was reprinted in 1997, a testament to the renewed interest in von Humboldt's scientific concepts and methods today.[1] The universal scope of von Humboldt's enterprise is almost inconceivable for a scholar today, but it represents a significant moment in a long intellectual tradition in which geography is embedded. That tradition of envisioning the material world as a *cosmos* has, as von Humboldt recognized, been continuously associated with the use of graphic images. Indeed von Humboldt's mapping innovations were themselves fundamental to a new approach to scientific method and communication in the nineteenth century.[2] The specialized term given to 'mapping' the cosmos is 'cosmography', little used in geography today but of very considerable interest to students of our discipline's culture and history. Cosmographers mapped spaces well beyond the surface of the earth, recognizing the inseparability of terrestrial and celestial forms and patterns. I argue in this chapter that there are growing reasons to revisit the cosmographic tradition within geography, but to do so from a contemporary cultural perspective rather than that of von Humboldt's 'physical description'.

To explain the interest cultural geography might have today in the spaces beyond Earth's immediate surface and atmosphere merits a re-examination of the historically deep connections between geography and cosmography. These were broken with the framing of modern science in the Enlightenment and during the nineteenth century. Von Humboldt's natural theology, like John Ruskin's landscape writing, clearly registers the tensions that attended that break. Philosophical and epistemological shifts in our own time have, however, weakened the tightly defined and policed disciplinary boundaries of modern scientific disciplines, and the universal claims and formal distinctions from subjectivity, creative art and imagination that characterized the Modernist scientific project. At the same time there is a growing human presence (virtual and actual) in spaces beyond the earth. Both these trends nudge us towards rethinking the cosmographic connection. They allow us, perhaps, to re-imagine a human geography of celestial space, a cosmography for the twenty-first century. I don't propose to sketch such a cosmography here, but to acknowledge that the actualities of a virtual world of informatics, remote sensing and the Internet, together with the realities of space exploration, provide evidence that an extra-terrestrial human geography already exists. It demands the attention of cultural geographers.[3] By drawing on well-established connections with the contemporary humanities and creative arts, cultural geographers are well positioned to address imaginative, aesthetic and poetic aspects of human presence in newly opened extra-terrestrial spaces, and to connect these with cosmography's long-standing mapping and graphic practices.

The chapter is in three parts: I first develop in greater detail the meanings of cosmography outlined in the previous chapter, and explore its relations to geographical and astronomical knowledge. For this I shall draw upon the account offered by von Humboldt himself. The second volume of *Cosmos* is a history of cosmography in which the author calls particular attention to the cosmos's metaphysical and artistic connections, while emphasizing their subordination to empirical observation in the years of European encounter with global geography. The second and third parts of the chapter echo in some measure von Humboldt's interest in the meeting of empiricism and imagination during the age of 'oceanic discoveries' and 'discoveries in the celestial spaces'.[4] While a historical sketch of cosmography is beyond both the space available and my own capacity, I want to recall in the second part of the chapter the crisis of cosmographic knowledge in the Renaissance, the consequent separation of Earth from the heavens, and the responses to these within humanist study and the creative arts. In the final section, I signal some indications of renewed interest in the poetics of cosmography and suggest how they warrant the attention of cultural geographers today, as extra-terrestrial space itself takes on a more complex human geography.

Cosmography

In his introduction to Volume I of *Cosmos*, von Humboldt discusses the meanings of the word cosmos itself, remarking on the shared sense of 'ornament' in ancient Greek that is still echoed in our modern word 'cosmetic'. Cosmos is a thing of formal order and beauty. Pythagoreans viewed creation as a harmonious and beautiful unity, whose order and proportion is most perfectly expressed in the mathematics of its form and motion. Ordered unity in physical creation remains von Humboldt's special concern:

> I use the word Cosmos in conformity with the Hellenic usage of the term subsequently to the time of Pythagoras. . . . It is the assemblage of all things in heaven and earth, the universality of created things constituting the perceptible world. If scientific terms had not long been diverted from their true verbal signification, the present work ought rather to have borne the title *Cosmography*, divided into *Uranography* and *Geography*.[5]

Cosmography's object is thus the 'order of the world'. As his words indicate, by von Humboldt's time cosmography had largely disappeared as a respectable scientific term, and in the succeeding years, the singular and universalistic implications of an 'order of the world' have rendered it a deeply distrusted concept in the social sciences and humanities too. From von Humboldt's perspective, Newtonian science and Enlightenment secularism had relegated the metaphysical concerns of cosmography to quasi-theological and popular devotional works, based on the remnants of an outmoded science, still seeking to justify a global providential plan in creation, and Christian salvation. Such natural theology found late defenders in writers such as John Ruskin whose early writings appeared in the same years as *Cosmos*.[6] Von Humboldt is concerned to reassure his readers that he is limiting himself to 'the domain of empirical data', and avoiding 'conceptions of the universe based solely on reason, and the principles of speculative philosophy'.[7] He recognized that such philosophical speculation had dominated cosmographic thought until the challenge to medieval scholasticism posed by modern exploration and experimentation. But if von Humboldt embraced a characteristic Enlightenment belief in positive scientific progress moving from myth to enlightenment by way of empirical observation within a disenchanted physical universe, he shared with Ruskin and other nineteenth-century students of nature the fascination with natural form and phenomenological apperception articulated by Johann Wolfgang Goethe.[8] Von Humboldt acknowledged the enduring power of human imagination and speculation in making and framing science. The final paragraph of the introduction to *Cosmos* states:

> the abuse of thought, and the false track it too often pursues, ought not to sanction an opinion derogatory to intellect . . . It would be a denial of the dignity of human nature and the relative importance of the faculties with which we are endowed, were we to condemn at one time austere reason engaged in investigating causes and their mutual connections, and at another that exercise of the imagination which prompts and excites discoveries by its creative process.[9]

Thus, von Humboldt separates physical science from cultural and historical studies, to the extent of devoting distinct volumes of his work to each. Volume I of *Cosmos* is a delineation of physical nature: a survey of then current knowledge of material phenomena in the terrestrial and celestial portions of the cosmos. Volume II is a history of the poetic, artistic and physical contemplation of the universe and of its scientific exploration, in which von Humboldt studiously avoids any critical consideration of what had long been cosmography's central moral question: the place of human order within the order of nature. Indeed, the structure of the work prevents this, reflecting von Humboldt's adherence to the strict division of objective and subjective knowledge (*Naturwissenschaft* and *Geisteswissenschaft*) that dominated nineteenth-century German scholarship. I return to this matter below. A more immediately obvious absence in *Cosmos* is illustration. Von Humboldt's text lacks maps or diagrams, conventionally key features of the cosmographic tradition. This is particularly notable given the author's own role in the history of thematic cartography: this is evident for example, in his famous description of altitudinal vegetation zones on Mount Chimborazo (see *Fig. 12.1*), and in the equally influential 1817 use of the isoline to denote a theoretical geographic surface produced by connecting empirically observed points of equal value, in a map of temperature variation across the northern hemisphere (see *Fig. 9.2*). That map demonstrated dramatically the inadequacy of describing global climatic patterns according to the ancient Greek *klimata* and zones of habitability. The *klimata* had been taken as evidence of cosmography's root assumption: that the order of the celestial world was inscribed on the elemental earth. Von Humboldts' had intended a separate volume of maps and illustrations to be prepared by Heinrich Berghaus to illustrate the argument in *Cosmos*, but the two projects became divorced and were published separately, to the detriment of each.

It can be argued that the graphic void in *Cosmos* points, as do the 1817 isotherm map, the work's division into two volumes, and von Humboldt's comment on changing scientific terminology, to the shrivelling of the cosmographic project in the face of the modern structure of knowledge and the disenchantment with the cosmos. As a mapping project, cosmography's conventional embrace incorporated and extended those of geography and chorography – the scales of the globe and its particular regions respectively – to figuring the whole of physical creation and proceeding onwards to its metaphysical template. Ptolemy's *Geography* had proclaimed mathematics as the technical foundation of geographic mapping of the earth's surface and its major units, with drawing and painting as the techniques of chorography, whose goal was to capture the character of place. Cosmographic mapping, although not specifically discussed by Ptolemy, called more fully upon the resources of imagination and speculation, with strong connections via geometry and number to Platonic ideal forms.[10] Identifying at the planetary scale those principles of order and harmony that render the Pythagorean universe a cosmos depends upon transferring observed temporal regularities (solar, planetary and stellar) to geographic patterns, and representing these graphically by means of points, lines and symbols: mapping as recording. But cosmography also called upon a more projective sense of mapping in its quest to calculate and predict consequential regularities that make the earth itself an ordered place of human dwelling: for example, Aristotle's 55 homocentric celestial spheres, belief in the symmetry of geographical patterns of land and water, or formal homologies between earth and the human microcosm.[11]

Cosmography's primary mapping act is the inscription of the 'heavens' (classically described in Aristotle's *De Caelo*) onto the elemental sphere composed of earth, water, air and fire (described in his *Physics, On Generation and Corruption* and *Meteorology*). This geocentric mapping is illustrated by the armillary sphere. Cosmographic terminology still structures our basic geographical vocabulary: horizon, axis and poles, equator, tropics and ecliptic (see *Fig. 1.1*). Aristotelian physics dictated that the geometrical perfection of the incorruptible celestial spheres, while inscribed theoretically on the terrestrial globe, for example in the bands of the *klimata*, could never be perfectly reproduced in the mutable spaces of the elemental world. The precise extent of variation from perfect symmetry and harmony remained a matter for empirical description. Thus emerged the enduring scientific questions of cosmography, such as the extent and balance of water and land on the earth's surface, or whether the northern *oikoumene*'s observed pattern of continental land, centred on a temperate zone, watered by the Mediterranean (Middle) sea and encircled by the Ocean stream, was paralleled in the southern hemisphere (see *Fig. 1.3*). How were such imperfections as the surface distribution of lands and waters, the atmospheric confusion of air and fire, or the corrugations that mountain ranges introduced to the earthly sphere to be explained?[12] Cosmography traditionally divided into a theoretical-mathematical side, which described the motions of the heavens and mapped their patterns onto the elemental globe, and a descriptive-graphic side, which outlined and described empirically the plenitude and variety of the global surface.

A principal purpose of cosmography in a pre-secular age was ethical: it concerned the place of human life in an ordered creation. It is worth recalling that the Greek word *ethos* referred to a 'sojourn', a familiar place haunted by a divinity (the Latin equivalent is *genius loci*). *Ethos* united earth and heavens, and implied the recognition that place and duration (habitation) inform and are reciprocally informed by conduct.[13] Thus, to know the time and space that govern the earth as a human habitation is to reflect upon how to conduct ourselves within it. In this respect, von Humboldt's division of *Naturwissenschaft* and *Geisteswissenschaft* stands counter to this traditionally governing principle of cosmos and cosmography. Plato devotes the final section of *Timaeus*, his influential cosmological text, to a description of the creation of humans from the materials of a cosmos, itself endowed with soul. The form and spirit of the cosmos are mapped into our material bodies:

> God gave the sovereign part of the human soul to be the divinity of each one [of three souls], being that which, as we say, dwells at the top of the body, and, inasmuch as we are a plant not of an earthly kind but of a heavenly growth, raises us from earth to our kindred who are in heaven. And in this we say truly; for the divine power suspended the head and root of us from that place where the generation of the soul first began, and thus made the whole body upright.[14]

This principle, slightly differently expressed in Aristotle's *Metaphysics* (*Lambda*) and *De Anima*, that we humans while sojourning on Earth have our origins elsewhere in the cosmos, gives cosmography an inescapable ethical dimension, deeply engaged with the moral and aesthetic questions of humanistic science.

Locating humans as the connecting centre of the cosmic map made cosmography doubly geocentric. Methodologically, cosmography proceeds from a central earthly sphere around which the planets and stars revolve. This remains true whether or not geocentricity is accepted as a scientific fact. Epistemologically, cosmography maps from the dual perspective of a human observer located on the earth's surface and a disembodied, divine eye located outside the cosmos altogether. Such a homocentric perspective lends support to the idea of a cosmos consciously designed as the stage for human existence, which is why the motif of the world as theatre remained persuasive into the early modern world.[15] The embrace of Greek science in the Christian West, especially after the twelfth century, saw the form and patterns of cosmos taken as evidence of the providence of a loving God to his principal creature and primary occupant of earth. Divine providence found its fullest expression in the microcosmic perfection of the redeeming Christ, at once God and Man.[16] Christian cosmography therefore performed the important ethical task of reading the book of nature alongside scripture as equivalent moral texts and witnesses to divine providence.[17] Early descriptive works such as Pomponius Mela's first-century CE *Cosmographiae* or Isidore of Spain's seventh-century *Etymologiae* and medieval mathematical texts such as Albert Magnus's *De Caelo et Mundo* or Sacrobosco's *Tractatus de Sphera*, employed as instructional works in European schools and universities well into the seventeenth century, mapped the order and variety of a universe centred on the earthly globe and its human occupants.[18]

Modernity and the crisis of cosmography

Modern thought came to reject both these forms of geocentricity, with consequences that are apparent in the structure of von Humboldt's Volume II of *Cosmos*. His 'subjective' account of cosmography – 'from the sphere of objects to that of sensations' – is itself subdivided into two parts, which he distinguishes both historiographically and textually. Volume II offers an account of stimuli to the aesthetic contemplation of nature, tracing, as we have seen, 'its image, reflected in the mind of man, at one time filling the dreamy land of physical myths with forms of grace and beauty, and at another developing the noble germ of artistic creations'.[19] This account summarizes a cross-cultural heritage of poetic and artistic descriptions of the natural world, and implies a widely shared connection between the human spirit and the material order of nature: an aesthetic, but not necessarily ethical, imperative within cosmography. The writing bears the strong influence of German Romanticism's attempt to re-enchant the secular nature inherited from the Enlightenment. Indeed, von Humboldt specifically addresses Goethe's complaint that in its accumulation of factual detail and failure of synthesis, German scientific writing served to make science inaccessible. Von Humboldt seeks to resolve any tension between his accounts with the suggestion that 'at periods characterised by general mental cultivation, the severer forms of science and the more delicate emanations of fancy have reciprocally striven to infuse their spirit into one another'.[20] Yet these 'delicate emanations of fancy' carry no special moral charge for him. The most significant of these periods when 'science' and 'fancy' seem mutually to infuse one another, and to which von Humboldt devotes one-third of the volume, comprises the quarter millennium between the Latin translation of Ptolemy's *Geography* (initially titled *Cosmographia*) about 1410, and the publication of Newton's *Principia* of 1670, a period when cosmography occupied the centre of European intellectual concerns.[21]

The tension apparent in Volume II of *Cosmos*, between 'scientific' and 'poetic' cosmography, is a characteristic product of von Humboldt's Enlightenment vision. His rejection of a humanistic cosmos is apparent in his treatment of Johannes Kepler, a pivotal figure in the evolution of a 'poetic' as much as a scientific cosmography:

> The figurative and poetical myths of the Pythagorean and Platonic pictures of the universe, changeable as the fancy from which they emanated, may still be traced partially reflected in Kepler; but while they warmed and cheered his often saddened spirit, they never turned him aside from his earnest course, the goal of which he reached in the memorable night of the fifteenth of May 1618.[22]

The 'memorable night' witnessed Kepler's discovery of elliptical movement of the planets, which, together with Galileo's observations of the Jovian moons, and more particularly the Italian's telescopic discovery of lunar relief and blemishes on the surface of the sun, radically undermined the Aristotelian assumption of celestial perfection in form and movement.[23] Von Humboldt's reference to the myths that warmed but did not charm Kepler's scientific spirit, is to Kepler's strong embrace of Renaissance Platonism, apparent for example in his attempt to relate the Platonic solids to the widths of the planetary orbs. But von Humboldt's response to Kepler's central spiritual dilemma is inadequate. The astronomer's whole project was ethical. As Kepler's 'Lunar Dream' (*Somnium seu de astronomia lunarii*) of 1634 amply demonstrates, his cosmographic images explored the earthly 'sojourn' of a body rooted in the heavens.[24]

In fact, von Humboldt's entire historical account in *Cosmos* underemphasizes the profound significance of Platonism within the Western cosmographic tradition. Yet Marsilio Ficino's translation of Plato's work and related Platonic and Neoplatonic texts for Lorenzo di Medici, including *Timaeus*, took place in the same city and in the same mid-fifteenth-century years (indeed among the same group of scholars) that reconnected Ptolemy's newly translated *Geography* with empirical knowledge of an expanding oceanic globe.[25] A direct example of the common context for mapping and Platonic philosophy in fifteenth-century Florence is Francesco Berlinghieri's rendering of Ptolemy's *Geography* into poetic Italian. Berlinghieri was a member of Ficino's Florentine Academy, and Ficino himself wrote the dedication.[26] While few would challenge the significance of the empiricism that von Humboldt's celebrates in generating a crisis within Western cosmography, historians of science today are far less confident than von Humboldt in distinguishing between poetic and scientific 'spirits' inspiring scientific revolution.[27] The argument over Copernicanism, for example, remained finely balanced well into the seventeenth century.

I have claimed that the cosmographic theorem of celestial perfection inscribed mathematically onto an elemental earth left open the question of how far the balance and harmony of the heavens were physically present across the elemental sphere. Mathematical cosmography long remained unaffected by heliocentrism, so that Apian's *Cosmographia* (see Fig. 1.1) and even Sacrobosco's *De Sphaera* were still being reprinted and used as instructional manuals a century after the 1543 publication of *De Revolutionibus Orbium Celestium*. Apian's cosmographic diagrams, for example, are reproduced on Antonio Campi's 1576 map of Cremona province to illustrate graphically the connections between local chorography, global geography and cosmographic scales of observation, while diagrams of solar and lunar eclipses from Sacrobosco are to be found in the margins of seventeenth-century Dutch and French world maps. By contrast, descriptive cosmography faced the much earlier challenge of oceanic discovery, and the rapid and widespread diffusion in texts and images of news about geographical distributions and the world's variety. From the early years of the sixteenth century the cosmographer would be faced with myriad and increasingly disconnected, even contradictory, fragments of personal observation, reportage, mappings and speculation arriving in his study from navigators whose journeys circled the globe. The consequences for the cosmographic project were predictable. Cosmographic texts became increasingly profuse and unwieldy, as Sebastian Münster's succeeding editions demonstrate, while the French cosmographer André Thevet's *Cosmographie Universelle* (1575) collapsed into incoherence, unable to sustain within a single controlling structure a descriptive unity for the earthly sphere to parallel that of the heavens.[28] In Frank Lestringant's words:

> The crisis of cosmography at the end of the Renaissance was manifested . . . on three planes. From the religious point of view, the cosmographer who raised himself to the level of the Creator in order to attain the latter's eternal and ubiquitous knowledge was guilty of pride, even blasphemy: he pretended to correct Scripture in the name of his sovereign, unlimited experience. At the level of method, he sinned by incoherence, confusing scales of representation and imagining that autopsy (or seeing for oneself) could guarantee the truth of a synthetic, and necessarily secondary, vision. Finally, from the epistemological point of view cosmography, which supposes a monumental compilation under the controlled authority of a single individual, was soon transcended by more supple and open forms of geographical knowledge.[29]

This early modern crisis of cosmography was primarily one of vision and representation, but it had profound ethical implications. Autopsy or eyewitness vision, guaranteed by either registering human presence (for example within the frame of a map) or by mechanization (for example by means of the camera obscura or optical lens, such as Galileo's burning sunspots through the telescope onto paper), was unavailable to the cosmographer.[30] He lacked a position from which to witness the truths he proclaimed. In fact, those who came closest to achieving such a cosmographic autopsy were not the scholars but the painters. A dramatic, if minor innovation in sixteenth-century Western art was the world landscape. The names of Joachim Patinir, Lucas Cranach, Albrecht Altdorfer and Peter Bruegel the Elder are curiously absent from von Humboldt's account of artistic achievements in the representation of nature. Yet the genre of painting which they pioneered came closer than any other to realizing Erasmus' cosmographic question: 'what spectacle can be more splendid than the sight of this world?'. Not only did these artists work in the same cities that many of the new techniques of mapping were being developed – Nuremberg, Augsburg, Antwerp – but a number of them had close personal connections with cosmographers: Altdorfer with Martin Behaim, Regiomontanus and Pirckheimer, Peter Bruegel with Abraham Ortelius.[31] Their often tiny, jewel-like panels were regarded by such scholars as a more adequate format than language or text for describing the universe, and indeed such images were actually referred to as 'cosmographies'. Albrecht Dürer wrote for example that 'the measurement of the earth, the waters, and the stars has come to be understood through painting'.[32]

In the years of Magellan's circumnavigation (1518–22) the panel painting was being held up by humanists as a paragon format for the description of the universe, and regarded as much more adequate than written accounts. A work such as Altdorfer's *Battle of Issus* thus maps the whole eastern Mediterranean as the setting for the global drama of Alexander's defeat of Darius's Persian army, raising the eye to a position where the site of the battle, the coasts of the Levant, the curving earth and the planetary bodies are all brought within its scope. Bruegel achieves a similar feat, mapping Crete and Cyprus into the cosmic frame of his *Fall of Icarus* (Plate 1) By relocating the eye in that liminal space between elemental and celestial spheres (where Plato's *Timaeus* located the human creature), these 'cosmographies' seem to anticipate the challenge to geocentrism while offering the eyewitness autopsy that evaded the textual cosmographers. And, while their themes may be imperial and heroic, the images are 'ethical' in the cosmographic sense in which I am using the term. Their genre quality offers to every observer the dignity and authority over space once reserved for gods and monarchs.

Extra-terrestrial geography

That the artists of the sixteenth century should have succeeded where the more scholastic cosmographers failed has a certain contemporary resonance. In all but the simplest respects, understanding today's cosmological theory is restricted to a small number of highly sophisticated mathematical physicists (although the success of Stephen Hawking's book indicates an enduring popular fascination with the subject).[33] Like most non-specialists, I can grasp only at the most superficial level such ideas as space-time relativity or multi-dimensional geometry, and understand such phenomena as 'black holes' or 'quarks' more as images and metaphors than as coherent concepts. I am told that fractal geometry offers new possibilities for establishing morphological correspondence between the unimaginably large and distant phenomena of the macrocosm and the most infinitesimally small particles of matter, but I do not clearly understand how fractals are derived and manipulated. Lacking such understanding, it is difficult to know in what sense, if any, the universe so described remains indeed a cosmos in the traditional sense of a harmoniously ordered unity. I hear promises of a single equation as the ultimate (and supposedly none too distant) solution to the questions of cosmic origins and form, but doubt if such an equation would be meaningful to me. However, when cosmological theories *are* rendered in ordinary language rather than equations, the reappearance of very ancient metaphors is often striking:

> The Aristotelian 'perfect cosmological principle', for example, of a universe maintained indefinitely by natural laws foreshadows the 'steady state' concept, whereas the opposed view of a 'big bang' theory of the universe starting from a single point and erupting out from there had its adherents [Epicurus and Lucretius] then as now.[34]

So, perhaps my conceptual inadequacies are not so disabling, and I begin to wonder whether it may not be that there is actually a strictly limited stock of representational metaphors and images upon which we are ever able to bring cosmos − the ornament of order − into meaningful being. Acknowledgement of the limits that representation places on all scientific knowledge has of course been a major philosophical advance of recent decades, a recognition that has demolished von Humboldt's neat distinction of objectivity and subjectivity, while revitalizing issues he regarded as superseded in a modern world. In the words of one recent writer on cosmology:

> Is the human race fundamental or incidental to the whole? A question that was once shunted to theology becomes increasingly relevant as we are made more aware that cosmology itself is a construct of human intelligence, subject to social and linguistic conditioning and dubious means of communication.[35]

As Plato long ago recognized, the concept of cosmos actually *requires* myth, symbol, graphic image − in a word, cosmography − to realise its ethical imperative. Introducing his spindle image of the universe in *The Republic*, Plato comments: 'to attempt to tell all of this without a visible representation of the celestial system would be labour in vain.'[36] The proclaimed 'golden age of cosmology' in the 1990s was itself in some measure a product of technical developments in visual imagery that have come with digital computing and graphics, and it has generated in its turn a renewed interest among creative artists in cosmographic themes. The stunningly beautiful coloured photographs of the cosmos with which we are now familiar are produced by processes of colour filtering and enhancement which are at once highly technical and deeply artistic. The photographer David Malin, who developed many of them, points out that the images are not visible in the eyepiece of any telescope. At best, even the most colourful gaseous nebulae seem little more than faint, luminous smudges of light, almost indistinguishable from galaxies of stars.[37] Malin's artistry yields results which are not too dissimilar to those created by the contemporary London painter, Adam Gray, whose oil paintings re-imagine the opposing scalar infinities of macrocosm and microcosm as the context in which organic life holds a special if disturbing place (Fig. 2.1 and Plate 2).

While we may have decentred the human creature from the contemporary cosmos, images such as Gray's reflect a continuing apprehension that organic life more generally holds a special place in existence; its discovery elsewhere in the universe – 'extra-terrestrial' – remains the most powerful stimulus to space exploration. And increased consciousness of both the constraints of representation and the temporality of humanity's sojourn on a planet whose own time within the cosmos is limited, intensifies rather than abolishes the ethical questions of how human existence should be

Figure 2.1 Adam Gray, Microbial

45

conducted. As humans, we remain quartered, as Heidegger put it, between earth and heavens, gods and mortals.[38]

But the ethical significance of the connection between earth and heaven is not confined to metaphysics. While theoretical cosmology remains at the margins of intellectual grasp for most of us, a human presence in the celestial spaces beyond the elemental sphere is becoming part and parcel of daily life. And, rather as the early modern extension of Europeans' presence into terrestrial spaces about which they had formerly only speculated generated new and pressing problems of 'human' geography, so the humanizing of planetary space today provokes more than simply questions of natural science. The impact of the first eyewitness views of earth from space, recorded in the Apollo lunar project photographs, resonated through late twentieth-century debates about the ethics of human life on earth.[39] At the turn of the millennium, the American artist Michael Light returned to the Jet Propulsion Laboratory's image archive of those extraordinary lunar missions between 1968 and 1972 when a dozen moonwalkers actualized Kepler's lunar dream. Treating as artworks images taken under the most self-consciously rigorous 'scientific' conditions, with a photographic eye sharpened in the desiccated landscapes of the American Southwest, Light's *Full Moon* maps an uncannily familiar extraterrestrial landscape (*Plate* 3). The lunar surface takes on a *human* geography, emerging as a landscape marked by human signifiers and familiar through the graphic conventions of landscape and cartography (suggested by the hairline grid that divides the photographic space), yet uncanny and disturbing because so many conventions and associations of terrestrial geography are here either stripped away or unaccountably absent.[40]

In these lunar landscapes, the presence and absence of light generate a metaphysical intensity. In the artist's own words, the images 'share a kind of delineation through distilled light that is at once highly abstract and yet brutally representational'.[41] Lunar light offers a sense of 'divine perception', and the astronauts themselves recorded their own apprehension of a god-like sublimity in its glow. In the absence of atmosphere, light and shadow etch across the Moon's monochrome surface landscapes with such crystal clarity and incised intensity that colour itself becomes the symbol of a softer, gentler Other. Items that in the context of earthly landscape might appear coldly metallic or artifactual – the struts or wheels of the lunar rover for example, or a power cable – gain in these photos unexpected warmth. And in one of the most moving images of Light's collection, the blue aura of water vapour surrounding Alan Bean's tiny figure as he stands against the lunar horizon, reminds us of ancient heroes – Jason or Ulysses – concealed by their protecting divinity who 'pours out' air (*pneuma* is the Greek word for spirit) from his body (*Plate* 3). This image serves too as an intimate reprise of the most haunting distinction created by the Apollo photographs:

between the Moon as the embodiment of the cold severity of celestial space and Earth as the watery, blue home of organic life. A tiny figure, trembling on the edge of deep space, the breathing human body becomes truly a microcosm and measure of life in the cosmos.

But physical human presence is not a prerequisite for an extra-terrestrial human geography. Ironically, much of the focus within human geography here on earth no longer connects necessarily to the material, mappable spaces of the physical environment, but works with the virtual spaces of imagination or of social interaction and connectivity, and with networks that do not necessarily depend upon the presence of material human bodies. In this respect, we might note in passing how many of the geographic networks we study extend into extra-terrestrial space, encompassing the dense landscape of satellites that now marks the innermost celestial sphere. Places and landscapes are no longer thought of by geographers simply as bounded containers, but as constellations of connections that form, reform and disperse in space and over time. Dwelling or inhabiting space is as much imaginative and conceptual as it is visceral and sensual. Such a perspective suggests that extra-terrestrial space does not have to be physically occupied in order to fall within the domain of cultural geography. And as planetary and celestial space beyond the earth's surface emerges as a realm of increasingly detailed human understanding and care, so a more complex and significant human geography is configured within it. From Soviet and American unmanned exploration projects of the past half century there is a presence of human artefacts in the atmospheres or on the surfaces of planets throughout the solar system. More significantly, their differentiated surfaces are increasingly present in the consciousness of men and women on Earth. We might suggest that such spaces are as present to us as were the sea-lanes and coasts of the great oceans to Europeans in the early modern world. And we should remember that the dramatic colours that render the deep space images such as those taken by the Hubble space telescope so aesthetically powerful are as much the product of human imagination and artistic skill as of cold science.

Conclusion

Cultural geography devotes much energy to exploring the meanings of place and human experience, of dwelling, attachment and rootedness. The tensions between these themes and those of mobility and cosmopolitanism, and the consciousness of belonging to a more diverse, complex and global space, created and sustained by social connections, are central to the theoretical concerns of cultural geographers in a globalizing world.[42] In this critical context, even the physical body of the human microcosm has been drawn within the scope of geographical consideration, reworked as a site of social and psychic construction and contestation, rather than merely a locus of material, organic life, shared with the rest of nature.[43] Yet for all its critical sophistication, the attention of cultural geography has tended to remain fixed at the surface of the earthly sphere. Its concern with socially constructed 'spatialities' risks detachment from those 'ethical' materialities that bind our embodied selves to the whole of organic life, and our spirits to spaces beyond the earthly surface, long the principal concern of cosmography. But, as the work of contemporary artists remind us, and as their Renaissance predecessors knew well, to lose the sense of the heavens is to risk losing also that of earth. Both spheres are inseparably connected to human existence. As twenty-first century experience actualizes and materializes this enduring truth in ways that are historically new, and as we create and live within an increasingly extra-terrestrial human geography, so cosmography as a cultural mapping, that aspect with which von Humboldt struggled, may find a revived significance. As von Humboldt himself pointed out:

> a sense of longing binds still faster the links which, in accordance with the supreme laws of our being, connect the material with the ideal world, and animates the mysterious relation existing between that which the mind receives from without, and that which it reflects from its own depths to the external world.[44]

II Landscape visions: Europe

3 Gardening the Renaissance world

The *Fioretti*, or *Little Flowers of Francis of Assisi*, is a medieval devotional text for followers of St Francis of Assisi. One tale describes the moment during the saint's 40-day fast on Mount Verna in Piedmont when Francis receives the stigmata. He stands at the entrance to his cell.

> Considering the form of the mountain and marvelling at the exceeding great clefts and caverns in the mighty rocks, he betook himself to prayer and it was revealed to him that those clefts . . . had been miraculously made at the hour of the passion of Christ, when, according to the gospel, the rocks were rent asunder. And this, God willed, should manifestly appear on Mount Verna because there the passion of Our Lord Jesus Christ was to be renewed through love and pity in the soul of St Francis.[1]

The moment is captured in Giovanni Bellini's small landscape painting made about 1480, now hanging in New York's Frick Gallery (Fig. 3.1). The artist captures the moment when St Francis receives the stigmata, the marks of the crucifixion, on his hands and feet, and sets it in an everyday landscape that, however luminous and beautifully realized, is recognizably based on the spaces of everyday life in the Venetian mainland of north-east Italy. Bellini makes a powerful connection between the inscription of divine love in the beauty of the natural world (which St Francis preached) and the marks of suffering and redemption on the saint's body. Viewing the landscape, the eye passes the evidence of the miracle almost unnoticed, and is led across a shelf of rock, through riverine pasture lands, to the enclosed and cultivated fields surrounding a fortified city. Its hilltop presence preserves order and due proportion in this humanized world. As spectators, we are firmly located outside the spaces represented, given neither a divine perspective – the divine rays that pierce the saint's body come from the upper left of the picture space – nor a distanced

Figure 3.1 Giovanni Bellini, *St Francis in Ecstasy*, 1480–85 (Frick Gallery, New York)

terrestrial position: St Francis already stands in prayer and fasting at the wilderness edge of the humanized landscape. We might almost see ourselves as explorers from another world, penetrating a pictorial space from a complementary angle to the rays of heavenly light penetrating St Francis's hands, feet and ribs.

Bellini's landscape is organized according to a social, gendered, corporeal and cultural hierarchy that was widely recognized in the Renaissance, and which dates back to Aristotle. It passes, as I have noted, from uncultivated wilderness, through domesticated pasture and farmland, to the rational, intellectual spaces of the city. Conventionally, the middle landscape within this hierarchy is the garden, a humanized space between wilderness and city. Were we to penetrate to the core of Bellini's landscape, we might imagine ourselves entering that hill town.

The town is surrounded by a thick wall, with towers and block houses at frequent intervals. On three sides of it there's also a moat, which contains no water, but is very broad and deep, and constructed by thorn bush entanglement. The streets are well designed, both for traffic and for protection against the wind . . . The fronts of the houses are separated by a twenty-foot carriageway. Behind them is a large garden, also as long as the street itself, and completely enclosed by the backs of other streets. Each house has a front door leading into the street and a back door leading into the garden . . . They're extremely fond of these gardens, in which they grow fruit, including grapes, as well as grass and flowers.[2]

This description, including its choice of garden cultivation, which we shall encounter more than once in this discussion, dates from 1516, some 35 years after Bellini's painting. It is of Aircastle, the most central of the 54 cities set on Thomas More's perfectly engineered but imaginary island of Utopia. Each of Utopia's geometrically planned towns lies at the heart of its cultivated territory, a perfectly arranged landscape for a perfectly ordered body politic.

These two descriptions set the scene for a discussion of landscape's embodiment, engineering and imagination in early modern culture. My guiding metaphor in this chapter is gardening: not in the sense of designing and constructing specific gardens (although garden designs were often conscious emblems of the currents of thought I shall discuss), but as a trope for Europeans' imaginative domestication of a new, global spatiality that was characterized above all by the disruption of previously established spatial, ethnographic and conceptual boundaries. And we know that fixing a boundary between the wild and the cultivated is the primary act of gardening.[3] To press the gardening metaphor, over the course of the European sixteenth century the *hortus conclusus* of Christendom was opened out and the wild world beyond threatened to transgress its broken paling. The Ocean Sea was not the only boundary to be crossed by early modern exploration. The sixteenth century also witnessed an internal exploration of the human body, not as the mechanical structure that it would later come to be, but as a geographical entity itself: 'the body was an (as yet) undiscovered territory, a location which demanded from its explorers skills which seemed analogous to those displayed by the heroic voyagers across the terrestrial globe.'[4]

Exploration, penetration and plantation, but also anxiety, disruption and corruption, characterized in equal measures both these fields of endeavour. Long associated with moral discourse, gardens and gardening offered an emblematic language for negotiating the spaces and landscapes of new worlds in the Renaissance. We might thus see the history of the garden as a dimension of Europe's larger imperial project, embracing the history of colonization in much more extensive ways than bringing back to cultivate in Europe the botanical curiosities, commercial and medicinal plants of new worlds, as was undertaken in the new botanical gardens established at Padua and Pisa in the 1560s, or cultivating herbal signatures in their anthropomorphically designed parterres.[5] As Europeans learned to navigate the oceans with relative ease, and as they came to recognize America as a distinct continental world, it was through metaphors of the garden and gardening that they imaginatively domesticated terrestrial and corporeal wildernesses. To focus my discussion I draw principally on Venetian examples. The justification for this is that although the lagoon city played a marginal role in actual discovery, its commercial and intellectual centrality in early modern Europe, together with its mapping and publishing traditions, made Venice the centre for collecting and disseminating information and ideas about the New World.

Landscapes embodied

According to Jonathan Sawday, out of Plato's division of flesh and spirit into distinct elements of human individuality, and out of theological belief in salvation through Christ's death and physical resurrection, late medieval Europeans had fashioned an understanding of the human frame as the site of a relentless war between body and soul. The body – fleshly, corruptible and gross, but also proportioned, beautiful and sensual – imprisoned and sought to impose its demands on the immaterial and immortal soul. Potentially capable of touching divinity, the soul was equally susceptible to being dragged down and shackled within the spiritual charnel house of bodily desire.[6] To achieve full humanity – and salvation – the rational and intellectual capacities of the soul had to be employed as tools to domesticate the wilderness of the body, and restrain its appetites and excesses, most notably in the realm of sexual desire. Not only was sexuality the most powerful of the body's uncouth urges, but its expression had complex affinities to love. And for Christians and Platonists alike, love was the moving power of divine creation, the principle to which the soul was ineffably drawn. Identifying and policing the boundaries between sacred and profane love was the most important and the most difficult of battles in the constant war between body and soul.

The garden was, perhaps unsurprisingly, the landscape in which this struggle was most dramatically located and embodied. Christ's passion begins in the Garden of Gethsemane as he struggles to assert a duty of divine love over corporeal fear of suffering and death. The tight enclosure of the medieval garden had protected virginal purity and maternal love from the predations of an unruly and wild, warring world beyond. But early Renaissance culture was reawakening pagan occupants of the garden: satyrs and herms, hybrid creatures, half-human, half-animal. Arcadians, Simon Schama reminds us, were conventionally autochthons, original men, sprung from the earth itself (Fig. 3.2).[7] Their most prominent physical

54

Figure 3.2 Coast Live Oak (*Quercus agrifolia*) grove, Arcadia California (photograph by the author)

feature was the phallus, unruly organ of male sexuality and instrument of insemination. The satyr embodied transgression of moral boundaries, the eruption of wildness within the pale, impossible to excise, but necessary to contain and bind. The moral struggle between soul and body, embodied within a garden landscape, is a constant theme in early sixteenth-century Venetian art and letters.

Conventional art history has often looked to the gradual exclusion of human presence and narrative as the measure of a maturing landscape art. In fact corporeal presence is central to understanding the cultural significance of landscape. Bellini's pupils Titian and Giorgione return again and again to this theme: the sacred virgin of the *hortus conclusus* has transmuted into the nubile Venus whose unclothed body reclines seductively in an unfenced landscape. The gendered invitation to observe, explore and penetrate is inescapable, but the order and fertility of the landscape recommends that self-restraint and cultivation rather than licence should characterize any bodily encounter within this space. In his eponymous painting of 1514 commissioned for the wedding of Niccolò Aurelio, the Venetian patrician whose coat of arms appears on the sarcophagus, Titian explicitly confronts the alternatives of sacred and profane love in the forms of a finely attired and worldly bride to the left, assisted by Cupid, and to the right a nude virgin holding the sacred flame of divine love. They are placed in a landscape in which it may not be too fanciful to read a dark and wild aspect behind worldly love, contrasted with the sun-filled pastoral scene behind her divine sister.[8] In sixteenth-century Venetian literature, the battle between body and soul took place in a gardened landscape: classically in Pietro Bembo's poem Gli Asolani (1505), composed by a patrician humanist in whose circle Titian moved. The work is a trio of Platonic verse dialogues on the nature of love, which take place between a group of young men and women in the palace gardens of the exiled Queen Cornaro of Cyprus at Asolo, a true locus amoenus in the foothills of the Venetian Alps. The garden is designed according to sacred geometry and replete with juniper hedges, laurels, vines and 'a glade of tender grass . . . edged by two groves of equal size, black with shade and reverent in their solitude'.[9] The moral conflict between Perotheno and Gismondo, representing physical desire and platonic love respectively, is ultimately resolved decisively, if a little unconvincingly, in favour of the latter.

But if the body in Bembo's garden resists earthly temptation, remaining pure and inviolable, the boundaries that protect it from the outside are not impermeable to the roving eye. Despite its enclosure, the garden positively invites a gaze over a broad plain beyond. From a height, says Bembo, a wide world may be seen beyond this garden. The landscape structure of Bellini's St Francis is here reversed. In Gli Asolani we gaze out from the centre into spaces of potential exploration. For Bembo and for many of his contemporary Venetians living through years of unprecedented change in Europeans' geographical consciousness and vision, the space of wilderness and exploration lay beyond Italy and the Old World: across the wilderness of oceans. The wild also lay hidden within the dark depths of the human body and of earth.

If Venice played little part in the Atlantic navigations, its government and citizens took an intense interest in them, and as one of Europe's principal centres of printing, Venice was something of a clearing house for information about the New World. Gianbattista Ramusio, Secretary to the Venetian Senate and friend of Pietro Bembo, published from the 1540s three volumes of Navigations and Voyages, one of the most comprehensive collections of exploration narratives of the sixteenth century.[10] These contained numerous and varied accounts of the landscapes and ethnography of the new lands. From Columbus himself, through Verazzano, Pigafetta, Cartier and others, narratives of exploration constantly swing between describing newly encountered lands as Edenic garden landscapes, their vines climbing through the trees, their parkland chases teeming with game, their gentle prelapsarian inhabitants supported by an indulgent earth under a mildly temperate climate; and limning them as howling wildernesses of tangled forest and swamp, haunted by terrifying creatures, half-man, half-beast.[11] The boundaries that contained such beastly monsters had once been as sharp as the frames that separate them from the inhabited oikoumene in Hartman Schedel's Nuremberg Chronicle of 1493. By mid-century such monsters had invaded the groves of arcadia, in their most terrifying incarnation as *anthropophagi*, eaters of human flesh.

In his 1580 essay on cannibals, the French humanist, Michel de Montaigne, pointed out that such creatures, if they exist, end up consuming themselves. Sawday points out the parallels between Montaigne's reading of *anthropophagi* and new ways of 'reading' the human body, then being opened to systematic exploration and public gaze in the new anatomy theatres. In its voracious hungers the human body itself came to be seen as cannibalistic. Sixteenth-century anatomy theatres, such as the one at Venice's University of Padua, were constructed in the 1540s on Vitruvian principles to represent the cosmic unity of human and earthly bodies. In the anatomy theatre, exploration of the anatomized body – a hugely popular public event – revealed how the body of earth in the form of food was processed and excreted through the human body, effectively an act of cannibalism. The title page to the Paduan Andrea Vesalius's *De Humani Corporis Fabrica* of 1543 places a woman's body at the central point of a circular theatre. She is as exposed to the roving gaze as one of Titian's or Giorgione's nudes (Fig. 3.3). Exploration has penetrated to the origins of life at the centre of corporeal space – the womb – breaching the boundaries of gestation itself and laying cruelly open the hidden core of the *hortus conclusus*.

In the same mid-century years Girolamo Fracastoro, doctor, graduate of Padua and friend of both Bembo and Ramusio, published his poem *Siphilides: Sive de Morbo Gallico* (1555), fixing forever the name for the disease which so

Figure 3.3 Giorgione, *Venus in a Landscape*, *c.*1501 (Gemaldegallerie Alte Meister, Dresden)

horribly disfigured the Renaissance body. Syphilis was the most terrifying evidence that the boundaries between the garden and the wilderness had been breached. The disease had appeared in the ports of Europe within months of Columbus's first voyage, and it ravaged the sixteenth-century world. Fracastoro's verse account of its origin takes place in an Arcadian landscape on Hispaniola, where a young shepherd, Siphilides, worships the earthly body of his beloved rather than the gods, and is struck down with the disease in punishment. Only when due sacrifice is made do the gods provide a sacred tree (*guanacum*) whose wood, the *lignum sanctum*, offers a cure. The complex relations between this narrative, the Edenic tree of wisdom, fertility myths of the golden bough, bodily sexuality and sacrifice, and the cultivation of nature, are too complex to unravel here. It is sufficient to recognize that Fracastoro's poem deals with another direct mark of exploration on the European body, a fearful consequence of breaking through the paling that enclosed the medieval world. As Fracastoro himself observed:

> Ganges and Indus soon were found to be
> No universal limits. Men were free
> To push beyond these barriers of the world,
> Towards Phoebus and his morning flag unfurled.[12]

But as syphilis indicated, a two-way traffic had been established across those broken barriers, and wilderness could erupt into cultivated Europe, into the European body itself. Such wilderness had to be tamed, and the response was plantation. Plantation meant precisely the erection of fences and the cultivation of wild nature. It meant imposing a gardened landscape across new worlds. In sixteenth-century terms it also meant the physical planting of living human bodies into the colonial earth. Among his late sixteenth-century writings, Francis Bacon includes an essay 'On plantations'. Plantations are a way, he says, for an ageing world to be renewed and to beget more children 'in pure soil'. And they are to begin as gardens within a landscape imagined almost as Edenic:

> In a country of plantation, first look about what kind of victual the country yields of itself, to hand: as chestnuts, walnuts, pine-apples, olives, dates, plums, cherries, wild honey and the like: and make use of them. Then consider, what victual or esculent things there are, which grow speedily, and within the year; as parsnips, carrots, turnips, onions, radish, artichokes of Jerusalem, maize and the like. For wheat, barley and oats, they ask too much labour: but with peas and beans, you may begin, and because they serve for meat as well as for bread.[13]

His advice continues, dealing with the morals of the men to be planted, and finally their relations with the 'savages' presently occupying the 'wilderness' surrounding the plantation. These last are to be used justly and graciously, but guardedly, sending some of them back to the colonizing country itself so that they may act on their return as ambassadors of cultivation. As the plantation 'grows to strength, then it is time to plant with women; as well as with men, that the plantation may spread into generations'.[14] In another essay, 'Of gardens', Bacon describes a remarkably similar scene, except on the smaller scale of 30 acres, divided into three parts: a green at the entrance, a heath 'framed as a natural wilderness' and the garden proper. This last is planted with those same fruits, nuts, vines and pineapples as he finds in the plantation landscape. The new world is to be variously domesticated with European bodies: it is to be an embodied garden set in wilderness.[15]

Landscapes engineered

In his text on syphilis, Fracastoro figures its spread to the European body as a moral consequence of the Spanish rape of Caribbean nature: European herms rather than planters in the new world garden. He compares the effects of the disease to the plague wind of late summer, when goats break out of their enclosures to engage in an orgy of wild mating before turning to a dance of death. In another image he compares the diffusion of the new disease within Europe to the River Po in flood, spreading uncontrollably across the fields. The recurrent trope is of a broken boundary, of free space across which the wild invades the cultivated. If plantation was the answer for new world wilderness, in Europe itself new boundaries had to be engineered to keep the wild at bay. And here again the outcome was to be an orderly garden landscape. We can observe the process in the territories immediately familiar to Bembo and Fracastoro, the wetlands around Padua and Venice where the Po's more northerly brethren, the Adige, Brenta, and Bacchiglione rivers, meander like swollen arteries towards the marshes and lagoons of the Adriatic Sea.

In one of his letters, the Venetian patrician, humanist and intellectual Daniele Barbaro describes the standard Aristotelian cosmology in a way that reflects the environmental specifics of the Venetian struggle between land and water, in which he was administratively and financially involved:

> Elemental nature is divided into four spheres, into fire, air, water and earth, and this is the constitution of the entire machine of the world . . . The nature of the four elements needed to be reconciled and moved by a superior virtue, and therefore celestial nature was introduced worthy, noble and far from contrarieties.[16]

The superior virtue of the celestial sphere is characterized by order and regularity, figured, as Barbaro's 1556 commentary on Vitruvius's *Ten Books of*

Architecture repeatedly insists, in the circle and by circular motion, or rotation.[17] Four quarters bounded by a circular enclosure constitutes the design basis for Padua's Botanical Garden, located near the anatomy theatre at Padua, in whose construction Daniele Barbaro was centrally involved. New world plants were first domesticated in this garden. Circular rotation regulates the great machine of the world, whether geocentric or heliocentric, and it is the fundamental principle of engineering, whereby human intellect and ingenuity bring the regularity of the celestial world to order the chaos of the elemental. In Barbaro's own words:

> The origin [of machines] derives from necessity, which moves men to accommodate themselves to their needs; nature teaches them and offers them examples either in animal life whence, it appears, many artifices have their origin, or in the continuous rotation of the world.[18]

The intellect was located in the head, the ruling or celestial part of the human microcosm. The machine extends the body's own capacity to intervene physically in the world, to reorder its elemental arrangement. In creating machines, humans thus mediate between elemental and celestial nature through the divine principle of rotation. In his commentary on Vitruvius, Barbaro illustrates this in a series of woodcuts showing mechanical devices, all of them combining mechanical rotation and human physical labour. Each is connected with water regulation and drainage, a matter of direct personal interest to Barbaro. With his brother Marc'Antonio he possessed a substantial estate in the northern Venetian plain, at Maser, close to Asolo. The property is best remembered today for the villa Palladio designed there and the *trompe l'oeil* landscapes painted on its walls by Paolo Veronese. But the Venetian archives contain records of the agricultural use of this estate, including the Barbaro brothers' cultivation of rice in large, reclaimed fields immediately south of the villa. Rice farming required both strict regulation of water systems and the management of large bodies of labourers, themselves reduced in the language of the day to mere body parts: *braccianti* (*braccia* = arms), whose diet, like that of the mid-sixteenth-century peasantry across the Veneto, was increasingly made up of cheap New World foodstuffs – maize and dried cod. The malarial disease so often associated with rice was one reason why the Venetians restricted the expansion of these unbounded expanses of stagnant water in the 1590s.[19]

Rice irrigation was but one element within a complex hydraulic landscape engineered across the Venetian plain during the course of the sixteenth century, a significant feature of what I have called elsewhere the Palladian landscape. Embanking and canalizing streams and rivers, constructing aqueducts and irrigation ditches, engineering locks and weirs to improve navigation, had all been undertaken in this region, as in many other parts of medieval Europe, since at least the twelfth century. But reclamation underwent a revolution in scale and engineering intensity from the sixteenth century. In 1556 the Venetians established a ministry devoted solely to the improvement of uncultivated lands, which in the Venetian plain meant essentially marsh and floodable lands. This agency worked closely with the Ministry for Waters, whose principal concern was to regulate and conserve the lagoon itself as a healthy space around the island city. The largest hydraulic works of the period were three great drainage projects undertaken in the lands between the Adige and Bacchiglione rivers south of Padua on the flanks of the Euganean Hills: the Monselice, Gorzon and Lozzo schemes. Describing the first of these, Nicolò Zeno the noble charged with supervising the drainage ministry, outlined its principles in terms very similar to Barbaro's:

> The Monselice scheme should proceed in three stages in imitation of the Lord God, who, in his creation of the world first separated the heavens from chaos [materia confusa], then separated the earth from the waters, and finally made the earth bear particular things: animals, trees, herbage. Thus we should conduct each scheme in three divisions.[20]

The outcome of such a cosmogony, generating new land according to God's own creative laws, would be a second Garden of Eden. Its boundaries would be defined by the survey lines, embankments, lines of tamarisk and poplar, and regulated channels through which waters that once had spread promiscuously across the land were to be contained and domesticated. The garden image is by no means unique to Zeno. In his submission to the Venetian Senate promoting projects for Zeno's Ministry for Uncultivated Lands, the humanist Alvise Cornaro anticipated the springtime landscape which would blossom across the region after drainage of lowland marshes:

> The laughing meadows full of sweet and varied flowers and suffused in smell; the laughing woodlands clothed again in bright young foliage; the laughing trees full of so many different and delicate fruits; the laughing vines releasing the sweetest odours from their blossoms; the laughing waters bubbling from clearer springs than ever before, their increased volume making a stronger murmur. So many different birds sing here, attracted by the fresh air: above all others the never-ceasing nightingale whose song is accompanied by the sweet notes of the cricket, the father of song. Singing, laughing, jumping, dancing and playing, the shepherds watch over their flocks grazing upon such nourishing herbage that they will produce the sweetest milk, so that to feed themselves the shepherds will have need of no other bread. All this singing, laughing and music-making comes from the new life brought to the hills by the liberation of which I have spoken, which has brought

them back to their original beauty, as they were when the divine
Petrarch chose to live and die amongst them.[21]

In a map of the Venetian Plain prepared for Zeno's ministry in 1556, the
year that Cornaro was writing, the surveyor and engineer Cristoforo Sorte
provided a pictorial image of this Venetian landscape (see Plate 4). It is a
bird's-eye perspective over the lowlands between the Alpine slopes and the
lagoon, at the centre of which lay the Barbaro estate. Colour coded to
identify drained and undrained lands, it pictures the territory as an intensively humanized landscape, a regular patchwork of enclosures surrounding
the city of Treviso. It could almost be an aerial perspective over the landscape Bellini drew upon for his St Francis. The aesthetic qualities of this image
should not surprise us, for Sorte had been trained by the mannerist painter
Giulio Romano, and later in life would write the first essay on the techniques of landscape painting. But his professional career was dominated by
engineering projects rather than painting (although he did decorative work
in the Doge's Palace at Venice). Many of his commissions involved surveying garden spaces and hydraulic engineering systems for patrician villas.
Among his colleagues in this localized work of landscape design were
Giacomo Gastaldi, close friend of both Gianbattista Ramusio, narrator of
oceanic discovery, and Girolamo Francastoro, writer on syphilis. Gastaldi
adopted the sobriquet cosmografo, although his official status as cosmographer to the Venetian Republic is questionable. But he did write a short
text on cosmography and is remembered today as the sixteenth century's
greatest Italian cartographer, producing some of the most detailed and
accurate maps of the newly disclosed world. The unenclosed blank spaces
on his maps are commonly pictured either as gardens occupied by innocent
Arcadians, or as the wilderness haunts of savages.

Landscapes imagined

In some senses all the landscapes I have been discussing were as much
imagined as material. The transoceanic lands and dissected bodies, which
most Europeans had never seen with their own eyes, were known to them
principally through the maps and graphic images printed in cosmographies
and the new atlases. Literary and artistic descriptions rendered Edenic
reclaimed lands, which in actuality were more often characterized by back-
breaking labour in the rice fields and malaria-racked human bodies than by
laughing fruit trees and contented shepherds. But, however idealized, these
landscapes did have real geographical referents. In this final section I turn to
purely speculative places within sixteenth-century Venetian landscape culture: landscapes whose geography and design lay beyond the reach of

human experience, either at the final ends of earth, in its inaccessible recesses, or in the unrecoverable past. In these, too, gardening played its role.

The group of Venetian friends that I have identified here, Pietro Bembo, Gianbattista Ramusio and Giacomo Gastaldi, also included Nicolò Zeno. They met and corresponded with each other, sharing common philosophical, scientific and cultural interests, gathering sometimes as an informal academy to discuss such matters. Their shared concern was cosmography: discovering, understanding and regulating an expanding world that was territorial, oceanic and corporeal. In his officially commissioned *History of Venice*, Pietro Bembo celebrated the recent revelation that the human habitability of the world was not restricted to the small part of the globe that the Ancients had known, and that within the New World Arcadian landscapes testifying to the earth's original state of perfection were to be found. On Cuba, for example, he claims: 'they lived in the Age of Gold: with no measure to the fields, neither judges nor laws, nor use of letters, neither commerce nor planning, but only living from day to day'.[22] But in the friends' private gatherings it was often more esoteric questions of cosmography that exercised their imaginations: the causes of the tides and their variation, terrestrial magnetism, the loss of a day experienced by circumnavigators, the internal form and nature of the earth's body. Among the most compelling of these mysteries was the River Nile, the subject of an intense debate between Fracastoro and Ramusio.

The Nile had been a geographic mystery for Europeans since antiquity (and would remain so until the late nineteenth century). It presented two related speculative questions: the cause of its annual flood and the location of its source. It is not pertinent here to discuss the alternative answers Ramusio and Fracastoro offered to these questions. More appropriate is to recognize that their interest in the Nile focused attention on a global geographic feature that directly linked humankind's earliest regulated and cultivated landscape with the most inaccessible of blank wilderness spaces on the world map. Above all, the Nile annually re-enacted the cosmogony itself, reducing the spaces of its delta floodplain to uncharted wilderness while at the same time covering them with such fertile mud that people could survive from the produce of garden plots. This was known through Vitruvius's account of the Egyptian origins of geometry in the annual astronomical resurvey of land parcels and gardens undertaken as the Nile waters retreated. Finally, the Nile itself was sometimes figured as a living body – in its fertility and indeed in its cannibalism – spawning a bestial life in its muddy depths in the form of flesh-eating crocodiles and hippopotami.[23]

During the sixteenth century, the Nile, its sources, and the ancient civilization it had produced, constituted a recurrent subject in the discourse and design of gardens. The four rivers that drained the world's continents: Nile,

Danube, Plata and Ganges, were linked to the four streams flowing out of the Terrestrial Paradise.[24] These have long structured the design of the garden in both Islamic and Christian cultures, just as they would structure a vision of universal Catholic space a century later in Gianlorenzo Bernini's great fountain in Rome's Piazza Navona. The ancient mosaic pavement at Palestrina representing the Nile's flood offered a rich source of iconography for European gardens well into the eighteenth century, principally through Francesco Colonna's hugely influential *Hypnerotomachia Poliphili*, published in Venice in 1499. The tale narrates the platonic pilgrimage of a young man towards Love and Enlightenment. Central to its argument is the dialectic of sacred and profane love, embodied in the landscape as a female body, by turns a source of pure, life-giving waters, and of sexual temptation attended by satyrs. The erotic woodcut illustrations in the Aldine edition of Colonna's work became a standard source for sixteenth-century garden design, and the book finds echoes in Pietro Bembo's *Gli Asolani* and in Titian and Giorgione's sleeping Venuses.

Imaginative landscapes from antiquity competed with the Nile as gardens of the world.[25] Another such landscape was Italy's coastal littoral between the Bay of Naples and Sicily, which contained gardened landscapes whose striking fertility owed their origin to the element of fire rather than water. The region is bounded by Vesuvius and Etna: the ancient world's two volcanoes. The Bay of Naples had of course been the favoured location for the villas and gardens described by Pliny the Elder and other Latin authors. Pietro Bembo's earliest written work, *De Aetna*, records his 1493 stay in Sicily and his ascent of Etna, prompted by the spirit of discovery that was even then directing Columbus across the Atlantic. His text is a dialogue in which Bembo, having returned to Venice, recounts his climb to his father as they sit together in the atrium of their villa overlooking the poplar fields and streams of the Padovano. The Mediterranean garden landscapes on the coast at Taormina are described in loving detail, and the richly fertile lower slopes of Etna claimed to be 'ever decorated with flowers and in continuous spring, so that it is easy for anyone to imagine that this indeed was the location of the rape of Persephone'.[26] As Bembo ascends the mountain however, he leaves behind the fields of Ceres and approaches Pluto's lair. This is howling wilderness, a landscape of sulphurous origins, violent winds, fire and primal chaos: 'clefts and caverns in the mighty rocks'. The remainder of the dialogue is devoted to speculation on the sources of volcanic fire and Etna's relationship to the whirlpools and tidal races of the Straits of Messina.

It is precisely this juxtaposition between the garden landscape at its most luxuriant and fertile, and the untrammelled, creative energy of the elemental volcano, that attracted the great northern humanist Abraham Ortelius and his companion, the botanical artist Joris Hoefnagel, to cross the Alps in 1573 and travel south from Rome and the gardens at Tivoli into the *bel paesaggio* of the Neapolitan Campania, the Bay of Naples and Sicily. The record of this early Grand Tour is a series of illustrations by Hoefnagel, in which the two companions point in awe at the wonders of what for them is at once an old and new world, a classical Arcadia and a place of chthonic power.[27] In 1573 Ortelius was flush with the success of his great atlas publishing project, the *Theatrum Orbis Terrarum* (1570), which offered a spectacle of the whole surface of the globe, across which the eye could wander while the body remained safely in the library or study (see *Fig. 9.1*). The metaphor of the atlas as a journey is sustained through the *Parergon* of ancient maps appended to the *Theatrum* in 1579. In its final pages, Ortelius comments,

> after this long and tedious peregrination over the whole world, I should bethink myself of some place of rest, where the painful students, faint and wearied in this long and wearisome journey, might recreate themselves; I presently, as soon as I awaked, went about it: and while I survey all the quarters of the huge globe of the Earth, behold the noble TEMPE, famous for their sacred groves.[28]

Ortelius borrows Colonna's conceit of dreaming in the garden as solace from a wearying journey in the world. Tempe, on the slopes of Mount Olympus, is a classical *locus amoenus* that the atlas-maker pictures as a gardened landscape of open meadow and woodland, watered by a river flowing out of Olympus (*Fig. 3.4*). Arcadian landscapes set on the islands and promontories of the Mediterranean Sea date to the earliest years of the Venetian Renaissance; it is not surprising, therefore, that Venetians should transplant these to the islands and shores of the Ocean Sea.[29]

One island landscape within the New World spoke with particular urgency to sixteenth-century Venetians and came to incorporate the contradictory features swirling through their collective geographical imagination. This was Tenochtitlan, the Aztec capital and the first great city encountered by Europeans in the New World. Cortez's letters describing the conquest of Mexico in dramatic prose appeared in numerous printed editions in 1524, accompanied with his map of Tenochtitlan (*Fig. 3.5*). The map was reproduced in Benedetto Bordone's *Isolario*, or island book, of 1528. The Venetian *isolario* tradition had long been associated with the classical ideals of the garden and *locus amoenus*. But as an island city, Tenochtitlan spoke very directly to Venetians at the very moment when their own iconographic tradition was building up the image of Venice as a perfect, insular world.[30] Jacopo de' Barbari's celebrated woodcut map of the city in 1500 pays detailed attention to the gardens or *delizie* stretching from the loggias of suburban villas on the Giudecca to the lagoon (see *Fig. 10.2a*). Tenochtitlan, set in the shadow of flaming volcanoes and surrounded by intensively cultivated and highly productive fields, laced with canals, and the centre of a

Figure 3.4 Abraham Ortelius, Tempe, *Parergon*, 1579 (Firenze, Giuni, 1991)

great empire, was directly comparable to the lagunar city.[31] Little wonder that on Venetian world maps the Aztec capital appears out of all proportion to its geographical scale. Yet the image of a garden-circled utopia was only one aspect of the city. Prominently marked at the centre of the city is the great ziggurat, a geometrical structure pictured like an ancient Egyptian pyramid, but devoted, chillingly, to the public dissection of the human body. Penetrate to the core of the New World landscape and one found, not Thomas More's peaceful city of communal gardens, but a brutal parody of the anatomy theatre. A more shocking alter ego to Venice, so often figured as the very model of an enclosed and protected *locus amoenus*, could scarcely be imagined. Little wonder that Tenochtitlan was so minutely pictured and described in Gianbattista Ramusio's published reports of the navigations.

Conclusion

Landscapes embodied, landscapes engineered and landscapes imagined: both Old World and New were being simultaneously discovered and gardened in diverse and complex ways over the course of the sixteenth century.

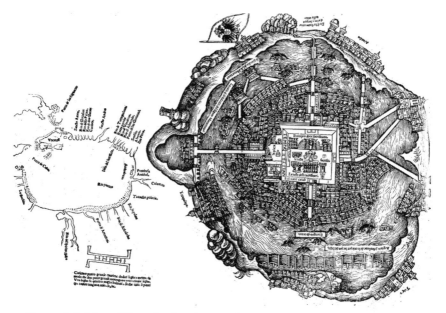

Figure 3.5 Map of Tenochtitlan (Hernando Cortes, *Praeclara Ferdinandi Cortesii de nova maris oceani Hyspania narratio* . . . Nuremberg: F. Peypus, 1524; University of Pennsylvania/Library of Congress)

Geographical discovery forced Europeans to respond to difference in complex and not always consistent ways, because transgressing boundaries always produces profound anxieties. Like the discovery of America, the exploration of the human body revealed a New World with fleshly appetites as raw as those of the transoceanic wilderness. Each threatened and attracted in equal measure; each demanded new responses, and for each, plantation seemed to offer a possible mode of domestication. In a world in which boundaries seemed to have been erased – where the pale was cracked and open – both new and very ancient forces could invade the European garden. New worlds meant boundless space. As Nicolò Zeno's words suggested, a restatement of divine language – the language of geometry, boundary lines, rotational movement and planting – might return us to the garden. But then again, Tenochtitlan and the anatomy theatre were constant reminders of a terrifying elemental chaos lying below the formal order of any *hortus mundi*.

4 Mapping Arcadia

The title 'Mapping Arcadia' signals perhaps the most clichéd assumption of the geographical project: making and interpreting maps. But as Arcadia is an imaginary place, the subject of artistic and literary more than geographic exploration, to map it suggests something more than surveying and compiling a graphic image of its dimensions and spaces. I use mapping here in its more metaphorical sense, but one that draws nonetheless on the deep historical connections between cartography and imagination. Such connections are apparent, for example, in the creation of patrician landed estates with their villas, vistas and gardens in sixteenth-century Italy, or in early maps of the New World, discussed in the last chapter. Here, I extend the connection between mapping and art, but use the idea of mapping as an active engagement that seeks to give form and meaning to an elusive and largely imaginary space.

Mapping and Arcadia

'Mapping', the cognitive and creative process rather than the scientific and design aspects of map-making, has more than one meaning. Most simply, it refers to the locating and way-finding practices of recording places and things in space, for example the charting activities traditionally employed by navigators that produced coastal *rutters* or *portolani*. But mapping also suggests more broadly cognitive and imaginative processes of discovering and denoting our place within the world, and of ordering the worlds we experience through spatial representations: graphically, pictorially, even narratively and performatively.[1] In the first sense mapping Arcadia would mean locating in geographical space an elusive region whose name has echoed through two millennia of literary and artistic culture as a point of harmony between nature and social life. Arcadia does exist as a geographical toponym on the map of Greece, a mountainous and wild region in the central Peloponnese. Its regional government today uses the name's resonance to promote eco- and agro-tourism in a landscape of poor hill farms and declining pastoralism. But the Greek region has only the most tenuous connection to the literary Arcadia, whose enduring literary and artistic influence is more accurately traced to Virgil's *Eclogues* and *Aeneid*. In Virgil, as I argue below, Arcadia might with equal textual justification be located in the reedy marshlands around Mantua in the plains of the River Po, the bare mountain pastures of Sicily, or the Capitoline Hill prior to the building of Rome: 'golden now, but once a-bristle with woody thickets' (*aurea nunc, olim silvestribus horrida dumis*).[2] And yet this classical mapping too is false to the literary tradition as we know it today, the origin of which is more appropriately found in the hugely influential poem *Arcadia*, written by the Neapolitan Jacopo Sannazaro in the 1480s and published (against its author's wishes) in Venice in 1502.[3] An immediate literary success across Europe, Sannazaro's poem celebrated the landscape of San Cipriano, near Salerno, south of Naples, where his aristocratic family had their estates.

Today, Sannazaro's poem is forgotten and unread – 'its preoccupation with beauty – beauty of countryside, beauty of language and of rhythm, idyllic beauty described in an elegiac tone', considered too saccharine for today's knowing and ironic tastes.[4] It comes to the contemporary reader mediated and relocated, through Sir Phillip Sidney's 'enamelled meads' and the rococo spaces of Marie-Antoinette's Versailles.[5] But at the time of its appearance, Sannazaro's poem was widely read and its landscape mapped onto the New World descriptions of sixteenth-century explorers. Here is Giovanni da Verazzano's record of the landscape he encountered near Cape Hatteras in 1524:

> following always the shore, which turned somewhat North, we came in a distance of fifty leagues to another land that appeared much more beautiful and was full of great woods, green and of various kinds of trees. Grapevines climbed to the branches, and the place we called *Arcadia*.[6]

Jacques Cartier expressed similar sentiments at Hochelaga, the site of today's Montreal, where he noted that people were engaged in cultivation and fishing solely to sustain themselves, and did not value worldly goods, being ignorant of them. Arcadia has been a recurrent theme in the European cultural appropriation of American nature. If I want to go horse-racing today in Los Angeles, or to the LA County Arboretum, I travel to Arcadia (see *Fig.* 3.2). It nestles below the San Gabriel Mountains just east of Pasadena, its appearance signalling the continuing power of poetic landscapes to become mapped into actual geographies.

Clearly, the broader sense of cognitive mapping is much more appropriate to the cartography of Arcadia than the narrow sense of finding our way there. For Arcadia's geography is one of yearning more than finding. It is a nowhere place, a utopia. Arcadia addresses the insistent question of the place that humans occupy, *should* occupy and, in reverie, perhaps once *did* occupy, in nature. Behind this single, simple toponym lies a complex geography of memory and desire, and a landscape that, once we map its poetic contours, reveals itself as the habitation of more troubling ghosts than we might initially expect.[7] This chapter attempts such a mapping by connecting some reflections on the current state of thinking about nature and landscape in North America to Arcadian thinking and its classical, Mediterranean origins, before returning to this now rather thin and tired trope of social harmony some of its complexity and poetic authority.

Arcadia in America

California, home of the Arcadia furthest removed geographically from those austere limestone crags and pine forests of peninsular Greece, is as good a place as any to start. The history of settlement in the Golden State over more than a century has been governed by a restless search for environmental and social perfection. The countless expressions of the California Dream across the Californian landscape – in sanatoria and health spas, on beaches and in communes, among orange groves and in the shade of eucalyptus trees, and even in the rows of suburban bungalows – make a story that is tragic in the original dramatic sense of that term: at once heroic and flawed.[8] It is shot through with the irony that the ideals of personal ease and social perfection sought within the setting of an idealized climate and spectacularly beautiful natural environment have been consistently elusive, endlessly deferred, or subverted in the artificiality of the human landscapes that have degraded Californian nature and hardened so much of Californian social and physical space.[9]

One aspect of this irony is captured in Roger Minick's image: *Woman with Scarf at Inspiration Point, Yosemite National Park* (1980) (Fig. 4.1). The Arcadian sweep of grassland, forest, rock and water visible from Inspiration Point appeals to a landscape sensibility shaped by Claude Lorrain's pictorial vision and moulded in the American West by Ansel Adams's photography[10] (Fig. 4.2). The woman's headscarf carries a banal printed image of the very scene she is viewing. The obvious effect of this piece is to question the 'naturalness' of both the scene she is observing and of her own response to it. In Dean MacCannell's phrase, Yosemite, paradigm of Western wilderness, is today 'Nature incorporated'.[11] Since 1855, Yosemite has been celebrated as the iconic landscape of the American West, the jewel in the crown of America's preservationist movement. Here the movement's founding father and guiding spirit, John Muir, played the Arcadian rustic, herding sheep in the

Figure 4.1 Roger Minick, *Woman with Scarf at Inspiration Point, Yosemite National Park,*
1980 (Oakland Museum of California)

1860s and proclaiming that 'nature like a fluid seems to drench and steep us
throughout, as the whole sky and the rocks and flowers are drenched with
spiritual life'.[12] Yosemite represents that uniquely American faith in
unspoiled, wild nature as the repository of natural (and national) virtue,
which, if embraced in the right spirit of subdued awe and reverence and in
accordance with the correct codes of conduct and dress, could transmit
itself to individuals.[13] Through the spiritual renewal offered by wilderness
experience, the moral order of society was supposed to be uplifted. This is
an Arcadian vision which spread across the globe during the twentieth
century. 'National' parks define 'true nature', the originating landscape, of
every nation state.[14] They signify, too, the nation's commitment to ideals of
environmental protection and conservation, to a balance between nature
and human life. Since the 1970s environmental science has been awarded
the privileged role of guide in the common human commitment to 'restor-
ing' harmony with nature (Fig. 1.4). In turn, ecology as the study of the

71

Figure 4.2 Ansel Adams, *Clearing Winter Storm, Yosemite National Park*, c.1935 (Ansel Adams Publishing Rights Trust)

balance of life has guided the management of national parks as the Arcadian groves of the modern world.[15]

But, over the past decade, the assumptions underlying the widespread commitment to wilderness and ecological virtue have been subject to a more critical scrutiny. The respected American environmental historian, William Cronon, initiated a fierce polemic when he pointed out in the mid-1990s that 'nature' is inescapably a human fabrication, a name through which we denote and delimit a reality that undoubtedly exists, with its own forms, processes and interactions, but which in and for itself is radically 'other', and thus, within human discourse, always a social construction.[16] The languages that we select to describe and address nature's forms and workings, together with the virtues or vices with which we invest 'wilderness', 'ecosystems' or any other of the terms we apply to natural spaces, as well as the roles we deem appropriate for ourselves in respect of them, are all culturally rather than scientifically determined. There is nothing pre-given or objective about the selections we make, and we need to attend critically to their implications. In thus promoting nature as a social construction, Cronon opened himself to the accusation of undermining the scientific foundations of environmentalism and opening the door to interests and activities that damage and destroy the natural world in the interests of profit or power. In fact, Cronon was simply asking for a more thoughtful, reflective and ultimately resilient conception of nature and how we should live in relation to the rest of life on this planet. Rather than asking natural science unaided to determine our moral choices, he urges us to relocate the conversation of conservation within the sphere of moral discourse, recognising nature for the human and social concern that it is. And that concern is one of the oldest and most consistent subjects of art and poetry. It is the key to mapping Arcadia.

Arcadia is precisely a place where the relationship between human society and the natural world is opened for critical reflection. To grasp the full richness and complexity of the Arcadian landscape, however, we need to escape the insipid images of courtiers and aristocrats playing pastoral games in Fragonard's world of the French *ancien régime*, or of world-weary city-dwellers going 'back to the land', to live 'naturally' in shapeless homespun on organic communes. We have to work our way back to the origins of the poetic trope and the contexts in which Arcadia emerged and evolved as an imaginative geography within the Western literary and artistic tradition. In doing so we might start with the designation of Yosemite itself. While not the first great tract of Western wilderness to be bounded and declared as a national park – that honour goes to Yellowstone – Yosemite offered the model as the earliest such space to be set aside, removed from the modern space and time that Americans prided themselves on making, to be appreciated for its timeless spiritual, scenic and recreational values alone. Although deep in the Sierra Nevada, by 1890 Yosemite was already accessible enough to the growing urban region of San Francisco to be exploited for both recreation and water – the fierce struggle over the Hetch Hetchy dam and aqueduct that tamed and took its waters was America's first great environmental battle. But what does it mean to designate such a place a park?[17]

Arcadia and the park

The origins of the park do not lie in the arrest of time or the exclusion of cultural impacts from nature. They lie in the enclosure of aristocratic hunting grounds, territories set apart from settlement and cultivation for the explicit purposes of the chase – hunting and killing wild animals. Areas such as Windsor or Sherwood in England, and Fontainebleau and Vaux-le-Vicomte in France, were woodland zones, their timber and game protected by law from exploitation by the surrounding community. Hunting lodges within or adjacent to them became palaces, villas and country houses whose immediate environs were transformed into gardens with elaborately designed orchards, flowerbeds and parterres. The 'wilderness' of the hunting park, which lay beyond the formal garden, provided for aristocratic leisure, to be sure, but it was leisure for a purpose. A society that relied for its defence upon an elite corps of mounted soldiers drawn from the nobility required space for the practice of martial skills: horsemanship and the use of heavy arms. The hunting park provided that space: it was the restricted-access military training zone of the feudal world. Its sylvan wilderness, a space of muscular challenge and violence, lay at the margins of aristocratic territory, beyond the space of cultivation and romantic assignation represented by the garden and the ruling centre of the country house, to be sure, but mapped securely within the social space of rule.

At first sight the conventional image of Arcadia seems far removed from the conceptual map I have drawn here of the park within the noble domain. In fact the two are closely connected. To understand the connection we first have to negotiate the Arcadian landscape sketched out for the modern world in the early Renaissance, the Arcadia later to be performed in the shallow aristocratic conceits of the Petit Trianon. The spiritual landscape of poetry and love where the genius of wild places and the simple life of nymphs and swains acts as a foil to the ennui of the city originates with Jacopo Sannazaro, the youthful Neapolitan noble and author of *Arcadia*. Sannazaro sublimated his adolescent passion in the poetic tale of Silvio, a jaded urbanite finding temporary solace in the woods and pastures of the family estates near Salerno. The enduring elements of the landscape sketched out in *Arcadia* are familiar: open woodland glades and pastures, frequent springs, streams and pools, and occasionally deep ravines and denser shades of gloomy forest, a landscape similar to that other mythical Greek scene pictured in Abraham Ortelius's image of *Tempe* on the lower slopes of Mount Olympus and later refined by Claude Lorrain (see Fig. 3.4). The land is inhabited by rustics: shepherds or goatherds and their animals. Autochthons, born of the soil, their movements are unplanned and unmapped, governed by the wanderings of their flocks, and their homes are either caves and temporary woodland shelters, or simple huts of branches, moss and leaves. The landscape is idyllic, pictured always in the full verdure of springtime, its dark sylvan recesses reassuringly distant from the sunlight shafting through oaken groves or warming the florid pastures. The Arcadians' rough manners and vulgar work are softened by their simple delights of rustic competition and their devotion to the poetry and song inspired by Arcadia's ruling genius, Pan:

> Nor are the shades of the trees . . . so discourteous that with their shade they altogether forbid the rays of the sun to enter the pleasant little grove; so graciously do they admit them that rare is that tree that does

not receive from them the greatest invigoration: and though it be at all times a pleasant spot, in the flowery spring more than in all the rest of the year it is most pleasing. In this so lovely a place the shepherds with their flocks will often gather together from the surrounding hills and exercise themselves there in various strenuous contests, such as hurling the heavy stake, shooting with bows at a target, and making proof of their skills in light leaping and stout wrestling, full of rustic trickery; and most often in singing and in playing the shepherds pipe in rivalry one with another, not without praise and reward for the victor.[18]

In Sannazaro's tale, various elements cloud this vision of simple but quite muscular harmony between man, beast, landscape and season. Most familiar among them are the sighs of the lovelorn young shepherd. In Sannazaro's lyric, the youthful rustic Ergasto's love is an affliction, and his personal melancholy shadows the landscape itself:

Spring and her days do not return for me,
   nor do I find herbs or flowers that profit me;
      but only thorns and splinters that lacerate the heart.
     . . .
How can you wish that my prostrate heart should rise
   to fix its cares on my poor and humble flock,
      since rather I fear it will be scattered among the wolves?
Amid my griefs I find no other remedy
   Than that of sitting alone at the foot of a maple,
      Of a beech, of a pine, or of a cork tree.
For thinking of her who has lacerated my heart
   I become an icicle and care for no other thing,
      Nor feel the pain from which I grow lean and waste away.[19]

Ergasto is only one of various lovesick youths who populate Sannazaro's *Arcadia*. Their adolescent yearning for a release from heartache leads them to a sacred grove 'which never did any dare to enter with any axe or iron', in whose craggy depths lies an image of the woodland god Pan. His altar bears the 'antique laws and rules for conduct of the pastoral life, from which all that which is done in the woods today had its first origin', and above it hang the waxen pipes. From this sacred grove the shepherds are directed to another, wilder place: 'a very deep ravine, bounded on every side by solitary and echoing forests of an unheard-of wildness; so beautiful, so marvellous and strange, that at first sight it strikes with unwonted terror the minds of those that enter there.' The band must pass through this sublime space in order to learn the rites that will either rid the youths of lovesickness or gain for them knowledge of the spells and potions that will 'force your enemy into loving you'.[20]

It is on their return to the brighter pastures of their flocks that the shepherds, about to burst into song, come across the pyramid tomb of Ergasto's mother, set between springs of water and against a grove of cypresses. The scene is the subject of the paradigmatic image of Arcadia, the seventeenth-century landscape painter Nicholas Poussin's *Et in Arcadia Ego*. The shepherds trace out 'the fitting epitaph' inscribed on the tomb (although Sannazaro does not actually record it): 'I too am in Arcadia.' Death, and the transience of youth, are central to the elegiac mood of the pastoral genre. In his classic essay on this image, Erwin Panofsky explored the complexities of Poussin's painting and the shifting connections between love, youth, mortality and the seasonal passage of time.[21] In his analysis, the inscription of death in Arcadia speaks of the subjection of all human life in the passage of natural time and the place of elegy as a poetic response to that universal experience.

The dark side of Arcadia

But for the modern reader there are more disturbing features than elegy alone to be found within Sannazaro's pastoral conceit of adolescent, urban melancholia toying with the pleasures and pains of a more 'natural' life. The testosterone-fuelled aspects of Arcadia have already appeared in the muscularity of the shepherds' recreations and the reference to the objects of their affections as 'enemies' to be 'forced' to return the young men's love. The pastoral pursuits of Sannazaro's Arcadia are insistently masculine, visceral and bloody. Here, the shepherds describe their hunt:

> beating the bushes now and then with clubs and stones we drove from cover thrushes, blackbirds and the rest toward the place where the net was. Fleeing before us fearfully, they inadvertently displayed their breasts to the deceitful snares, and entangled in them hung here and there as if in so many little sacks. But at last when we saw that our catch was enough, bit by bit we gave slack to the ends of the master-ropes, bringing them to the ground; where some bewailing their plight, some lying semi-conscious, they abounded in such plenty that many times grown weary of killing them and having no place where we could put up so many, we carried them along with us, all caught up in the ill-folded nets, to our accustomed abodes.[22]

Other passages describe in similarly graphic terms bear-baiting and torturing wounded dogs. Death has a sanguinary face in Arcadia.

To comprehend Arcadia's darker side more fully, we need to leave Naples for Rome, and to go back fifteen centuries from Sannazaro's time to the eve of the first Christian millennium. Sannazaro's landscape is in many respects a pale reflection, if not a complete distortion, of the Arcadia sketched by Virgil. Perhaps the most consistently influential poet in the Western canon, Virgil's three great works, the Eclogues, Georgics and Aeneid, collectively constitute both a history of cultural evolution from an initial state of nature to the refinements of urban civilization, and an epic celebration of the development of Rome from its founding to its apotheosis as Augustus's capital, Virgil's patron and Rome's first emperor. Arcadia appears only sporadically, in four of the ten Eclogues, and frequently then only as a passing reference within rhetorical poetic appeals to Pan as a god of music and rhyme. In a detailed examination of the poems, Richard Jenkyns points out that only one of the references connects Arcadia to the geographical region of Greece; the others seem to place it in Thessaly, Sicily and in Virgil's own home region of Mantua in northern Italy.[23] The name 'Arcadia' is evoked principally because of its association with the god Pan and with the Greek poetic genius more generally. The only Eclogue specifically set in Arcadia scarcely resembles the world of Sannazaro's nymphs and shepherds. The land is figured as a bleak and cold terrain in which there is a palpable sense of a way of life coming to an end: 'thy loved Lycoris through the inclement camp / and snowy march follows another flame'.[24] In all the Eclogues 'toil is one of those things – like women, like war, like winter – which are not in the foreground of the poems but are seen and heard at a distance'.[25] For Virgil, the pastoral world is not a place of escape from the city, nor indeed a meaningful geographical location, but rather a particular and passing historical moment in social evolution.

This becomes much clearer in the Aeneid, Virgil's epic of the founding of Rome. Having arrived on the shores of Italy after their passage from the destroyed Troy, Aeneas and his companions seek the support of a group of Arcadians in establishing their authority. The Arcadians themselves are exiles, removed from their Greek homeland and reduced to the extremes of a primitive existence on the scrubby hill that would in time become the Roman capitol. Virgil is providing Rome with a Greek heritage as much as a Latin one, mapping Greek nature onto the banks of the Tiber as he unites Aeneas's Trojans with Evander's autochthonous Greeks to create the Roman race. The home of the exiled Arcadians is a place of destiny:

> Even then the place's holiness awed the peasants;
> even then they trembled at the rock and forest.
> 'This grove' he says 'this hill with its crown of foliage,
> some god lives here – we do not know which one – my people
> think that they have seen almighty Jove himself
> shaking his dark aegis and gathering the storm clouds.[26]

Evander leads Aeneas to his rustic house, 'they saw cattle there, / bellowing in the Roman Forum . . .'.[27] The king's house is constructed of trees bent to form a roof, his bed is of leaves, its cover a bear skin.

We are in the realm of nature at the earliest moments of its domestication by human community. The image of Evander's rude hut poised at the junction of natural and cultural landscape is one that will resonate through the European imagination: we find it in Vitruvius's description of the origins of architecture, and among Sannazaro's contemporaries in the opening decades of the sixteenth century, where the scene is pictured along with the discovery of fire as the founding moment when humans differentiate themselves from the rest of the natural world by controlling and shaping their environment.[28] It is not surprising therefore that the natives of the New World, recorded for Europeans in precisely these same years, should have been pictured as golden age Arcadians. Ethnographic descriptions of Virginia, for example, owe as much to Virgilian description as to direct observation. In the Virgilian epic, Evander's son Pallas, Aeneas's closest friend, his ally and relative, dies at the hand of Turnus, leader of the native Latins and the Trojan leader's rival in love and war. The final lines of the *Aeneid* describe Aeneas's bloody vengeance as he refuses the wounded, defeated Turnus's pleas for mercy, and plunges a sword into his heart. Imperial Rome, born in the blood of native and colonizer, is here universalized though a combined Latin, Trojan and Arcadian pedigree that unites the origins and ends of history and geography.

The significance of this poetic construction becomes clear if we return for a moment to Evander's description of the earliest occupants of the land now occupied by the Arcadians and in future to be the city whose empire stretched 'to the ends of earth':

> Native Fauns and Nymphs once lived in these forests,
> and a race of men sprung from the trunks of sturdy oaks,
> who had no rules or customs, could neither yoke the ox
> nor lay up supplies, nor save what they had gained.
> But boughs of trees and the rough fare of hunters fed them.[29]

Jupiter descends from Olympus and gives these people law. He changes the name of the place to Latium and brings in the golden age. But that age does not last; it collapses into strife and war. Only Aeneas's own heroism and eventual death will produce the proper circumstances to guarantee the return of the gilded age, which Virgil implies is imminent in the imperial rule of Augustus.

The Lucretian narrative

More significant for my argument than the obvious ideological role of the *Aeneid* in the service of Augustus, is the source for Virgil's narrative of nature and culture among the Arcadians. It is the poet Lucretius, whose *De Rerum Natura* appeared when Virgil was in his twenties. Lucretius's work is an example of a form now wholly absent from our culture: the scientific treatise written in verse form, whose authority as science rests as much in its rhetorical forms and poetic structure as in its conceptual and empirical contents. Lucretius offered a dramatically different vision of nature from the more familiar Aristotelian description of a mutable elemental world of earth, air, fire and water placed within a perfect and unchanging celestial realm. Lucretius's cosmos consists of atoms, 'preserved indefinitely by their absolute solidity'. It is a world without origin or end, but of continuous change and reformation. And the governing force that powers this mutation is sexual love. His poem begins thus:

> life-giving Venus, it is your doing that under the wheeling constella-
> tions of the sky all nature teems with life, both the sea that buoys up
> our ships and the earth that yields our food. Through you all living
> creatures are conceived and come forth to look upon the sunlight.
> Before you the winds flee, and at your coming the clouds forsake the
> sky. For you the inventive earth flings up sweet flowers. For you the
> ocean levels laugh, the sky is calmed and glows with diffuse radiance.
> When first the day puts on the aspect of spring, when in all its force
> the fertilizing breath of Zephyr is unleashed, then, great goddess, the
> birds of air give the first intimation of your entry; for yours is the
> power that has pierced the heart. Next, the cattle run wild, frisk
> through the lush pastures and swim in the swift-flowing streams.
> Spell-bound by your charms, they follow your lead with fierce desire.
> So throughout seas and uplands, rushing torrents, verdurous meadows
> and leafy shelters of the birds, into the breasts of one and all you instil
> alluring love, so that with passionate longing they reproduce their
> several breeds.[30]

Here is something much more vital and immediate than the languorous melancholy of Silvio's longings, although lust's presence in Arcadia is figured in the satyr and the Priapic statue. For Lucretius, there is nothing romantic in the principle of love: it is urgent and constant, necessary for the continued existence of elemental nature. But Venus neither does nor can act alone. She is but one half of the union that sustains universal life. Her lover is Mars, god of war and bringer of violent death and destruction. In the opening hymn to Venus just quoted, Lucretius begs her to tame Mars, to lay him low 'by the irremediable wound of love', in this case more a matter of the heart than the loins. Lucretius is offering more than a poetic conceit or a reference to the civil war raging in Rome at the time of the poem's composition; he is signifying the dual processes of procreation and destruction by which the atoms that make up the world are endlessly rearranged: 'the universal mother is also the common grave'.[31] In Lucretius's work the death that is in Arcadia is not necessarily the peaceful passing of age and decay; it arrives in the violence of rage and conflict. It resembles more the Darwinian struggle for survival than Sannazaro's sweet melancholy. It is a vision of nature hot in copulation, 'red in tooth and claw'.

For Lucretius, human society too must be subject to this elemental duality. In his sociology the originating human nature is not one of Edenic innocence and perfection, but rather one of brutality, violence and lust. The earliest humans lived 'in the fashion of wild beasts roaming at large':

> They could have no thought of the common good, no notion of the mutual restraint of morals and laws. The individual, taught only to live and fend for himself, carried off on his own account such prey as fortune brought him. Venus coupled the bodies of lovers in the greenwood. Mutual desire brought them together, or the male's mastering might and overriding lust, or a payment of acorns or arbutus berries or choice pears.[32]

For Lucretius, culture is a process of learning, not the gift of any divinity but a long process of trial and error, always threatened by a return to brutality: 'in struggling to gain the pinnacle of power they beset their own road with perils', sinking back more than once 'into the turbid depths of mob-rule'. Each discovery that extends human control over nature also extends the human capacity for violence and destruction. But the inventiveness of war in turn extends human authority over nature. 'So each particular development is brought gradually to the fore by the advance of time, and reason lifts it into the light of day.'[33]

Arcadia and imperial consolation

We seem perhaps to have drifted a long way from Arcadia. In fact we are very close to grasping the roots of that unique landscape. Virgil's Arcadia occupies a specific place in the narrative of imperial Rome. Drawing upon the authority of Lucretius's natural philosophy, Virgil uses his predecessor's cosmology and sociology to naturalize and universalize a national epic for Roman Italy. The Arcadians play a critical role in this, for they connect the origins of imperial Rome to the origins of nature and human society itself. They represent the first society, described by Lucretius thus:

As time went by, men began to build huts and to use skins and fire. Male and female learned to live together in stable union and to watch over their joint progeny. Then it was that humanity began to mellow. Thanks to fire, their chilly bodies could no longer endure the cold under the canopy of heaven. Venus subdued brute strength.[34]

But this is at best a temporary and fragile truce; Arcadian society is not destined to last. Not only is death in Arcadia, as everywhere, but violence and war echo constantly through the Arcadian glades. Mars cannot long be held in Venus's embrace, for his distinct contribution is fundamental to the continued evolution of social life. Thus in Virgil's epic Evander's son Pallas has to die, and the Arcadians, like the Latins, have to be swept away for their country to be colonized by the Trojans. Yet their sacred hill will become the very heart of Rome, and thereby in Roman ideology the centre of the world, and their blood will invigorate Trojan stock.

Thus, the idea that Arcadia and the pastoral originate in an 'urban literature, expressing the longing of city sophisticates for vanished simplicities . . . the happy lives of nymphs and swains, and sometimes satyrs, . . . a realm of poetry and love, a natural idyll, . . . a land of lost content'[35] is profoundly mistaken, attending to only one side of a more ambivalent and imperial story. In Virgil, to whose poetry it has conventionally been traced, Arcadia is not an imaginative or desired place of stasis or of achieved harmony between humans and the natural world; it is a moment within a complex process of human evolution whose driving forces are sexual love and violent death.

The Renaissance image of Arcadia is not wholly evacuated of its Lucretian origins. However vapid, the melancholic yearnings of the lovesick swains, contrasted to the Priapic lustiness of satyrs and of Pan himself, carry faint echoes of the evolution in sexual relations described by Lucretius. The presence of death in Arcadia is a banal reference to human participation in the cycle of nature, but the pyramidal stone tomb with its inscription signifies a warrior's monument. The competitive games that Sannazaro's Arcadian shepherds play retain a muscular and martial cast, and their brutal treatment of wild birds and beasts reminds us that the Arcadian landscape is the home of the hunt.

Yet it is the Virgilian rather than the Sannazaran Arcadia that more accurately anticipates the historical geography of those Arcadias that Verazzano and Jacques Cartier believed they had discovered in the New World. Their inhabitants may have been described as living in the golden age, in harmony with nature, in a simple community where 'mine' and 'thine' had no meaning. Historical and anthropological reconstruction of pre-Columbian societies reveals highly complex and stratified communities woven into sophisticated commercial networks. But the fate of these American

Arcadians, like that of Evander's people before them, was to be destroyed in the construction of empire, their landscape giving way to the cartographic logic of the colonizer: the straight military road and the rectangular fields that spread across both Roman Italy and modern North America. The difference in cultural response to these two moments is, however, profound. Virgil looks with pride over the achievement of empire, seeing the destruction of the Arcadians as a necessary stage in the progressive evolution of human society. He acknowledges that the price of empire is high, but the autochthons have contributed to its glory and they are immortalized in the legitimacy they give to the geography of imperial Rome. Its centre is their own sacred place and their blood roots the present inhabitants of the city in the natural landscape.

Like Aeneas's Trojans, America's colonizers frequently made (and broke) pacts with one group of native peoples against another, and the consequences there too ultimately entailed the destruction of both subjected peoples. But at the very moment of destruction, in the same years that the US armies bloodily defeated the last of the Plains Indians, the first Western American landscapes were being set aside as permanent Arcadias for the modern world, indeed sometimes on the very sites of battle. And in this case the designation was explicitly and resolutely cartographic, as the Act of March 1872 that designated Yellowstone National Park makes precisely clear:

> That the tract of land in the Territories of Montana and Wyoming, lying near the head-waters of the Yellowstone river, and described as follows, to wit, commencing at the junction of Gardiner's river with the Yellowstone river, and running east to the meridian passing ten miles to the eastward of the most eastern point of Yellowstone lake; thence south along said meridian to the parallel of latitude passing ten miles south of the most southern point of Yellowstone lake; thence west along said parallel to the meridian passing fifteen miles west of the most western point of Madison lake; thence north along said meridian to the latitude of the junction of the Yellowstone and Gardiner's rivers; thence east to the place of beginning, is hereby reserved and withdrawn from settlement, occupancy, or sale under the laws of the United States, and dedicated and set apart as a public park or pleasuring-ground for the benefit and enjoyment of the people; and all persons who shall locate or settle upon or occupy the same, or any part thereof, except as hereinafter provided, shall be considered trespassers and removed therefrom.[36]

Significantly, it was the US Army that policed Yellowstone and other national parks until the establishment of the National Park Service in 1915. While John Muir was discovering in Yosemite an American Arcadia in which to play shepherd and to meet his God, America's native peoples were engaged in the final death struggle for their independence. In this sense the national park and the Indian reservation are twin geographical expressions of the same process, enclosed by the same gridiron cartography and by the same colonizing authorities, a process that we recall today more in shame than in pride.

Conclusion

What does the 2000-year long story of Arcadia tell us about nature as an elusive paradise? One is that Arcadia is better understood as a landscape of consolation for the imperial bad conscience than for the heartaches of adolescence or the world-weariness of urban life. Another is that perhaps Lucretius and Virgil were more honest and true to the relations between humans and the natural world than Sannazaro and John Muir. Arcadia can only be located in time, never mapped into space. The attempt to fix geographical boundaries and thereby contain nature is as arbitrary as the grid lines confining the US national parks, and as doomed as contemporary attempts to contain the aspirations of native peoples within the same cartography. The element of fire is relevant here, for Lucretius placed great emphasis on mankind's control of fire in taming wild nature, in humans themselves as much as in the world around: it signalled the beginning of culture. Fire suppression was a central management strategy in twentieth-century attempts to preserve wilderness nature in the national parks, ignoring the historical evidence of native Americans' extensive use of controlled burning in their land use strategies, and the fact that much of the picturesque beauty of forest and meadowland in places such as Yosemite was actually the outcome of such practices. Fire suppression has proved one of the most dangerous and destructive management policies in the parks and is now abandoned in favour of selective and controlled burning. Excessive restrictions on hunting can have similar consequences for deer or mustang populations. As for the bounding of native peoples, all parties are shamed by a system that long treated them as if they truly were merely elements of a separate, natural world. Only now are the descendants of America's Arcadians recovering their own history, and they still await from the majority society the kind of respect for their contributions to the America epic that Virgil accorded the Arcadians. Perhaps they would not wish it.

Another lesson is that we should resist the sentimentality of Sannazaro's juvenile flirtation with a wholly romanticized nature, his embrace of what John Ruskin called the 'poetic fallacy': the conceit that nature itself responds to human feelings. Yet we should also be cautious not to erect impermeable barriers between the scientific understanding of the forms and processes of nature and our aesthetic responses to the natural world. By aesthetic here I mean the response that comes through the senses: the forms and colours, sounds and silences, smells and tastes of the world around us and our bodies' direct contact with it. A recognition of the common goals of art and science underpinned the long tradition of using poetry to map out natural philosophy. It may not be appropriate to resurrect that tradition, but today, as the realms of art and science long separated by Romanticism begin to reconverge and overlap, new maps of nature and new cognitive mappings of Arcadia become possible.

III Landscape visions: America

5 Measures of America

The American landscape makes sense from the air. This is not so true for more reclusive European landscapes, which level with the observer slowly, only as we enter into them. European visitors to North America, from John Winthrop to Hector St John de Crèvecoeur to Jean Baudrillard, have recorded in their different ways a common experience, increasingly apparent as one travels West across America, that here milieu gives way to space.[1] And Americans themselves, for example the painters Thomas Cole and Albert Bierstadt, or the poet Walt Whitman, have been awed by the vastness of America's spaces and the scale of the American nature their art sought to capture. The idea of the sublime that was applied to the experience of landscape in the eighteenth century, around the time of American independence, seems especially appropriate for New World scenery. In the era of manifest destiny, Whitman and the historian Frederick Jackson Turner went so far as to proclaim territorial space the very foundation of American nationhood.[2] In popular culture, freight train blues, highway ballads and the road movie still relay the romance of endless movement towards a specifically American horizon. The aeroplane, now the dominant mode of long-distance travel in America, has rendered visible the lineaments of American landscape in the everyday experience of American space.

There are deep historical and cultural reasons for the distinction between American landscape and European landscapes, and between the synoptic, aerial perspective by which American landscape is most clearly revealed, and the intimate, fragmented embrace appropriate to European locales. Those reasons are closely connected to ideas of measure. Here, I survey some of the historical and cultural forces that have measured and sculptured out of American space the broad features of landscape visible from the air.[3] Since about 1600, those forces have originated largely outside the continent, but their outcomes have been distinctly American. They have sought to inscribe beliefs, desires and visions, originally nurtured in the crowded, fragmented and bellicose little worlds of that splintered, sub-continental peninsula which is Europe onto what colonial eyes arrogantly took to be the *tabula rasa* of a new world. In the process of inscription, earlier cultural impressions on American nature – impressive earth mounds created by ancient communities along the upper Mississippi in Illinois, cleared fields in the eastern woodlands, Hopi villages and Navajo pueblos in the arid Southwest – have been set adrift within a refashioned landscape, reduced to relict, archaeological echoes of past modes of human existence. To most contemporary eyes they are largely of romantic appeal: ciphers of a cosmographic measure often believed to be absent in modern America.

Aerial visions

The aerial perspective is uniquely appropriate for imagining and understanding the American landscape. The twentieth century opened with the human conquest of the air in North Carolina in 1903 and with the insertion of Oklahoma, New Mexico and Arizona as the final pieces in the geometric jigsaw of the coterminous United States in 1907. The cultural and political implications of flight echoed through the century, among them flight's transformation of the ways landscapes are seen and interpreted.[4] In this chapter I shall highlight two seemingly contradictory, but closely related, responses to the aerial view of American landscape. Each concerns ways of measuring earth, expressions, we might say, of a geographical imagination. One concerns space, the other nature.

The aeroplane is the most visible of a great range of modern technologies that have progressively annihilated space by time over the course of the past century.[5] The frictional effects of distance, the time and energy expended in moving across space, so painfully apparent on sea and land, are dramatically reduced in flight. The boundaries that disrupt terrestrial movement and fragment terrestrial space disappear in flight, so that space is reduced to a network of points, intersecting lines and altitudinal planes. In his three-month celebratory tour of America in 1927 following the famous New York–Paris flight, Charles Lindberg covered 22,000 miles and 82 stops, touching every corner of the country.[6] Connections between disparate places become easy, even routine, with the consequence that place distinctions seem to fade, nowhere more apparently than in locations directly associated with air travel: airports, hotels or leisure resorts. The earth's topography itself flattens out to a canvas upon which the imagination can inscribe grandiose projects at an imperial scale. From the air, the imposition of political authority over space can be readily appreciated: the die-straight linearity of Roman military roads, the geometry of baroque urban plans, the gridded fields of reclaimed polder-lands, the sweeping masonry curves of

China's Great Wall. From the beginning, Modernist planners and architects were captivated by the aerial view. The dream of flight, offering an Apollonian perspective of the wide earth, encouraged visions of rational spatial order to be written across the land, free from the hindrance of local contingency and variation.[7] In this reading, the political and cultural impacts of human flight revolve around the capacity for synoptic vision, rational control, planning and spatial order. A continental scale nation such as the USA offered a practical as well as imaginative canvas for such schemes. It is not surprising therefore that the great visionary engineering projects of the New Deal, such as the Tennessee Valley Authority and the Western dams on the Colorado and Columbia rivers, were conceived in the early years of commercial flight and promoted through panoramic, often aerial, photographs.[8]

Alternatively, and sometimes simultaneously, the aerial view has encouraged a new sensitivity to the bonds that bind humanity to the natural world. Consider the architect and planner E. A. Gutkind's comments to a mid-century environmental symposium. Gutkind had been centrally involved in the physical reconstruction of Europe after each of its devastating twentieth-century wars. In 1956 he spoke at the Chicago conference entitled 'Man's Role in Changing the Face of the Earth', organized by the geographer Carl Sauer to assess environmental impacts and futures. Gutkind's subject was 'Our World from the Air'. 'Today', he claimed, 'we can look at the world with a God's-eye view, take in at a glance the infinite variety of environmental patterns spread over the earth, and appreciate their dynamic relationships.'[9] Gutkind believed that such a view encourages synthetic rather than analytic thought: 'everything falls into a true perspective – even man himself as an integral part of the whole'. With the help of nearly 50 aerial photographs, Gutkind drove home the central theme of the symposium: that humans are at once insignificant against the great measure of nature, and yet capable (if theirs remains the only measure) of wreaking the effective destruction of earth as a home for human – and possibly all – life. What Gutkind tried to demonstrate in his photographs was that the measure of nature is as much local and contingent as it is global, and that the aerial view must ultimately lead to a renewed appreciation of the organic and the local even as it is distanced from earth itself.

The ability that flight offers to see the world from above culminated in the 1972 Apollo photograph of the spherical Earth, whole and unshadowed[10] (see Fig. 1.4). Responses to that image have stressed both the unity of the global vision, and the need for local sensitivity in a globalized world. Late twentieth-century commentators echoed Gutkind in stressing that such a graphic demonstration of unity between humanity and nature should have lasting and positive effects on social and environmental thinking.

Nowhere have these two twentieth-century themes – of space annihilated by time, and of the localized ecological bonds of nature and human life being threatened by our own hubris – played themselves out more vocally than in the United States. In part, this is because of the pivotal economic, political and cultural roles that America has played in the world since 1945. But, arguably, that role has simply allowed long-standing American predilections about land and life to gain global traction. To understand those predilections we need to recognize the historical nature of Americans' social relations with space and American nature, and thus take the measure of American landscape.

Cosmographic measure

America occupied no space in either the charts or the mental maps of the Genoese, Spanish and Portuguese navigators who sailed out of the Mediterranean into the open Atlantic. Western and eastern shores of a single world island faced each other on the cosmographer's globe across an ocean of unmeasured breadth. North America became a continent in the European consciousness even more slowly than South America, and perhaps not fully until Cook entered the Bering Sea and Lewis and Clark reached the mouth of the Columbia river, little more that two centuries ago.[11] For its initial discoverers America was another group of islands in the Ocean Sea – like the Canaries, or Madeira and the Azores, which had emerged out of the Atlantic mists during the previous century. Yet these same Europeans had measured the globe's dimensions, theoretically since Eratosthenes, and would do so physically from Columbus's and certainly Magellan's navigations. America may have been blank space, but measured space awaited its appearance. Its absence was fixed on the graticule of astronomically determined lines of latitude and longitude.

That graticule represents an abstract, intellectual inscription of measure across the globe. It is a measure calculated by spherical geometry and coordinated with celestial movements. Although directed towards practical ends on earth, geometry had its origin in the heavens, and it presupposed a distanced, rational optic. Geometry's Platonic characteristics as a pure product of mind with immutable and therefore divine properties connect the history of its uses in cosmology and cosmography to the practicalities of latitude and longitude. The poetics of mathematical measure run through both.[12]

Heirs to the Renaissance reworking of classical knowledge, the New World's European discoverers were fired by its belief in the powers of geometry and mathematical measure. Their immediate predecessors had created new spaces of representation by applying perspective geometry to painting, working out the practicalities of spherical projection in map-making, and reviving the geometrical measures of ancient Greek and Roman architecture.[13] Geometry provided the Renaissance measure of Man, that self-fashioning invention of the years of discovery, whom Pico della Mirandola placed at the mid-point between the celestial and terrestrial, the divine and the bestial, and whom Leonardo da Vinci fitted perfectly into the circle and square.[14] Geometry united macrocosm and microcosm; it was the secret measure by which creation had been ordered and was sustained. It thus behoved God's highest creatures to employ that same measure in making their own, lesser worlds. Through geometrical measure Man would author the earth, and most especially a new found land to the West. The line of Tordesillas drawn in 1493 was the first act in a continuous intellectual measuring and shaping of America by European geometry. Sometimes hesitantly, often confidently, the outlines of a continent would be inscribed over the course of four centuries into the chequered frame of the planetary graticule, while the chequerboard itself was transferred onto American territory in the property lines of colonies, states, communities and individuals.

In the European fashioning of American space, cosmographic measure pre-dated and prefigured the encounter with actual *nature*. Colonists took the view that the Native American population was so much a *part* of nature (literally savages: 'of the woods') that its members were incapable of intellection. Even their most sympathetic observers deemed the disembodied, synoptic vision implied in the geometrical measure of European cartography beyond them. Captain John Smith claimed of the natives of Virginia:

> They are as ignorant of *Geography* as of other *Sciences*, and yet they draw the most exact Maps imaginable of the Countries they're acquainted with, for there's nothing wanting in them but the Longitude and Latitude of Places.[15]

We know now how mistaken such commentators were. The contrary evidence is most apparent in the ancient landscapes of the Southwest. Locations at Hopiland, or the circles and squares at Pueblo Bonito, were founded on equally precise astronomical measurement and resonate with similar mathematical poetics as those of ancient Attic temples or the courtyards at Renaissance Urbino. In designs that plot these pre-Columbian measures onto the rectangular grid of the United States Geodetic Survey (USGS) topographical map series, the contemporary landscape architect James Corner drives home this point (Fig. 5.1). For all these cultures, the measure of the skies denoted a cosmic finality whose relations with terrestrial nature demanded constant intellectual attention: if their measure is upset, nature itself is affected: 'the time is out of joint'. Thus Navajo and other people of the American Southwest made ritual use of elaborately geometrical sand paintings that are regarded by scholars as maps of their world.[16]

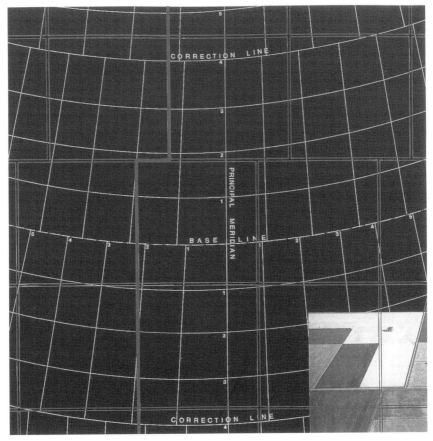

Figure 5.1 James Corner, Grid landscape design from the USA (James Corner and Alex MacLean, *Taking Measures: Across the American Landscape*, New Haven: Yale University Press, 1996)

Cartographic measure

As its European conquerors slowly grasped the scale of this 'new world', they searched for some mode of coming to terms with and controlling the immensity of American space, and the difference of American nature. Measuring and mapping, imagining the landscape in the mind, or inscribing it onto parchment or paper, are more rapid, less dangerous and more secure ways of coping with unknown spaces than penetrating the Appalachian forests with axe, or trekking the featureless grasslands of the Staked Plains of Texas. The map offers a way of controlling nature's contingencies or the rivalries of human proprietors. In the settlement of America, each colonizing community faced the joint needs to balance the freedoms and physical dangers offered by immeasurable space against the safety and social constraint offered by measure, rule and boundary. The sermons of New England Puritan divines repeatedly address the social contradictions generated by open and bounded spaces.[17]

Transferring the map from the study to the ground itself created landscape in the union of American nature and European imaginings. This process too has its own history, a history inflected by the cultural backgrounds of settlers and the environmental context in which they found themselves. In the dense forests of the Eastern seaboard and the lake and river lands of the Northeast, measure by the stars was rarely possible in any consistent manner. Little wonder then that the French, who first penetrated forested North America along the rivers – *coureurs du bois*, whose search for furs took them from their snowbound St Lawrence landfall through the Great Lakes, the Ohio, Missouri and Mississippi river systems – used the river banks as the base line for their cartographic surveys and property systems. Equally spaced straight lines, surveyed away from the banks of the St Lawrence or the Chaudière rivers, have produced *rangs*, or long lots, that form a ladder pattern of properties marked in the landscape by patterns of pasture, grain and woodlot. Plotted from a meandering river such as the lower Mississippi, the lines inevitably diverge or converge, producing fan-like property borders whose logic is apparent only from the map or the air. Such a cadaster may seem at first sight more responsive to the contingencies of nature than the astronomically determined patterns of the Spanish pueblo in the treeless Southwest, or the Anglo-Saxon rectangular grid across the prairies. But rivers, especially those that flow in the vast floodplains of the Ohio-Mississippi system, alter their courses with a regularity that leaves the long-lot landscapes of Louisiana frequently disconnected from the water frontage that determined their initial logic.

The scale of much American nature, like that of its rivers, was quite out of proportion to anything that European settlers had ever experienced. One of the very few words left to us by the obliterated native peoples of the Caribbean is *hurricane*: a natural phenomenon of such violence that Europeans lacked their own names for it. Despite the endeavours of European thinkers such as Count Buffon to represent American nature as degenerate (based on its paucity of large mammals), with the intention of elevating Europe to the peak of an imagined evolutionary hierarchy, the experience of those who actually confronted American nature was more frequently of its enormous scale and energy: the endlessness of its forests, the height and force of its waterfalls, the dimensions and fury of its storms, the numberless herds of buffalo and flocks of passenger pigeons which covered the plains and blotted out the sun.[18] The ability of American nature to overwhelm conventional European literary and artistic languages is nowhere more apparent than in the records of the Lewis and Clark expedition, sent out by Thomas Jefferson to describe the landscapes of the newly acquired West and to disprove such slights as Buffon's to American nature. As the explorers moved beyond the Great Falls of the Missouri into the true mountain West their written accounts, whose form Jefferson had so carefully tutored at Monticello in preparation for the expedition, fractured and failed.[19] Artists accompanying the railroad builders across the same spaces seven decades later experienced similar challenges to the graphic and painterly vocabularies they had learned in Europe and the East.[20]

Jefferson's role in the cartographic creation of American space was crucial, well beyond his purchase of French Louisiana, its largest single territorial block. Historically, movement of European settlement beyond the eastern forests onto the open prairie grasslands coincided with the young Republic's great democratic experiment to ensure for its people 'liberty, a farmyard wide'.[21] For the Founding Fathers, children of the European Enlightenment, and above all for Jefferson, mathematical measure was not so much the divine attribute it had been to America's Renaissance discoverers as the tool through which the moral and social goals of an enlightened new republic could be achieved. Rational monetary measure – dollars and cents rather than pounds, shilling and pence; rational weights and measures – the reformed gallon and the geographic mile; and rational division of land into rectangular states, townships and sections: all these were intended to create a new and more perfect society. And it was 'open' American space west of the Appalachians and the Ohio river that offered the unique historic opportunity to achieve a dream of utopia that had been planted in the European brain in the first years of oceanic navigation. Having devised in 1785 a method of surveying and selling unsettled space into individual ownership (the Land Ordnance Act), as president Jefferson set about purchasing, exploring and mapping a continent.

The rectangular survey system, whose square townships, die-straight property lines and field boundaries, and abrupt discontinuities, dominate the agrarian landscape of the Mid-West and the Plains, may appear – to our ecologically sensitive eyes – a hubristic expression of insensitivity to nature and topography, to landscape itself (Fig. 5.2). But we should remember its original intention within the context of a European world in which citizenship had always been founded upon ownership of immobile property, that is, upon land. Opening American space in equally sized parcels, at an affordable price, to individual farmers, appeared the precondition for a stable and open democracy. The rectangular grid is the perfect spatial expression of the new republic's democratic and localist imperatives. While the axially converging lines of the princely Renaissance city (Fig. 5.3) or the baroque garden are centre-enhancing, proclaiming power, absolutism and centralized authority, the rectangular grid is space-equalizing. It privileges

Figure 5.2 Alex MacLean, Grid landscape: Western USA (James Corner and Alex MacLean, *Taking Measures: Across the American Landscape*, New Haven, Yale University Press, 1996)

Figure 5.3 Anonymous artist, *Panel Showing an Ideal City*, c.1480 (Walters Art Museum, Baltimore)

no one point above any other; it distributes power equally across space. It is the landscape measure of America's commitment to life, liberty and the pursuit of happiness. And the vastness of this conception, corresponding in its universality to the Enlightenment belief in a single, rational order in nature, is a fundamental determinant of America as landscape rather than landscapes.

The foundational principle of land allocation enunciated by the Land

Ordinance Act of 1785 was that land should be surveyed and recorded on the plat before settlement. While this, like the strictures against speculation, was honoured more in the breach than the observance, it reinforced the cartographic foundation of American social space, laying the surveyor's measure across nature's local contingencies. The insistency of its geometric logic is most apparent on the USGS topographic map where the design conventions and choice of landscape elements to be represented privilege geometry over topography. The American topographic sheet is a network of intersecting lines and numbered squares within which the eye struggles to discern farms and schools, cemeteries and churches. No European topographic map is so fiercely linear and numerical. On the ground, matters are different. The grid confronts us in the straight road with its sudden dogleg turn, necessary to accommodate flat survey to curving earth, in the regularity of passing farmsteads when driving through Illinois or Ohio, but in little else. Only from the air does the global insistence of its controlling geometry reflect that of the map.

Natural measure and time

For nature too has its insistencies. From the air above the fields of Kansas, the dendritic geometry of watercourses strikes dramatically across the rectangular survey blocks. In Minnesota or northern Wisconsin, relict features of the geologic past interrupt the universality of Enlightenment landscape logic: drumlin fields, kettle holes and esker sand ridges that recall ice advances and retreats, measured according to a geological time scale. What is remarkable when seen from the air is that it is these 'natural' features that seem out of place within the landscape's logic, rather than the human measure seeming inappropriate in the context of the marks left by vast and ancient natural forces. On the ground, the relative roles are reversed, as the human landscape appears to submit to nature's prior and unchallengeable claims.

On the map such topographic features as glacial drumlins and pre-Columbian burial mounds are marked by contour lines, appearing as islands in a charted sea, whose curving, intricate patterns seem almost to subvert the logic of rectangular geometry imposed by the grid. As European settlement moved beyond the humid lands of the Missouri-Mississippi plains, out into the semi-arid and arid West, the tensions between nature's measures and those imposed by human minds sharpened. The West was conquered by railroads and barbed wire, rather than by heroic frontiersmen or pioneer wagon trains. Before the townships and ranges of the federal land surveyors could be turned into a chequerboard of fields and crops, the railroad engineers and navigators had to bolt lines of steel across the continent. Preceding them was of course the second continental survey (more accurately surveys) funded by the federal government. It took the form of expeditions and their multi-volume reports, illustrated by topographic sketches and finished drawings of Western landscape, that helped feed imperial visions of continental settlement among legislators, boosters, speculators and immigrants from every part of the Old World.

The railroad and its inseparable partner, the electric telegraph, more than any other technologies, were responsible for the annihilation of space by time in the nineteenth century. As elsewhere in the world, it was the timetable demands of rail transportation that prompted the unitary measure of time which overtook local solar calculation in 1883. America was remeasured into four time zones across its continental space. The great turning circles constructed at the railhead termini in Chicago or Minneapolis seem from the air like great clock faces, reflecting on the ground the timepieces that once dominated the stations themselves. The rail network often cuts uncompromisingly across the grain of the rectangular survey grid, and the towns and regularly spaced grain elevators that still line its tracks across much of rural America record the supremacy of an urban market over the Jeffersonian vision of the self-sufficient yeoman cultivator in the Western landscape.

Confidence that nature had been nailed down by geometry was shared equally by early railroad boosters in New York and Chicago and isolated homesteaders on their quarter or half-quarter sections dotted across the semi-arid plains of Oklahoma and the Texas panhandle, with only the railway line to measure an endless horizon. Such confidence was misplaced. Only the most sensitive adjustment to the vagaries of local nature will allow continued cultivation west of the hundredth meridian. Contour ploughing and other dry farming techniques have been the precondition of Plains agriculture since the environmental and social catastrophe of the Dust Bowl. Thus even in the most topographically featureless regions of the United States, where the linearity of the survey line and the railroad track reign supreme as the unchallenged measure of the landscape, nature indirectly reasserts an alternative in the serpentine contour line: the measure of level. The surveyed lines of natural topography followed by the plough and scarcely apparent on the ground are highlighted on soil conservators' maps and visible from the air as bands and colours of alternating cropland, weaving a more complex landscape texture across the warp of the rectangular survey. Flying into Kansas City from the West on an October evening reveals an abstract Impressionist landscape whose geometry combines Jackson Pollock and Mark Rothko: meandering lines of woodland picking out the branching watercourses and highlighted by the occasional flash as the sun catches their surface, the smooth curves of the contoured plough lines revealing the topography in warm browns of rich earth, green bands of pasture and the ochre tones of dried cornfields, and the whole landscape framed in the chequerboard pattern of the rectangular grid.

Modernist measures

The contoured landscape dates from the mid-twentieth century, from the years of the New Deal, when, for the first time since Jefferson's mastering vision for the continental landscape, the federal government became involved in large-scale landscape planning. And once again, survey and mapping preceded action. A vast archive of soil quality and watershed maps survives in Washington as testimony to America's third great continental survey. But the interwar planners had a technology at their disposal unavailable to nineteenth-century land and railroad surveyors: the aeroplane and the aerial photograph. These provided the technical support for the synoptic vision that is Modernist planning. A buzzword of mid-twentieth-century engineering, whether in aircraft design itself or in hydrological planning, was *streamline*.[22] The streamline implied more than just speed – the latest stage in the conquest of space by time: it also implied rational, scientific efficiency, achieved through adjustment to natural flows. Water or airflows around obstacles always follow the path of least resistance. Such flows suggested to planners a paradigm for human interventions in nature, allowing natural properties to be directed towards human ends, as in the case of flight itself, where air moving over the wing planes lifts the craft off the ground. Some saw affinities between the idea of the streamline and the Chinese cosmological concept of *feng shui* (literally wind-water), which acts within Chinese geomantic theory as an elemental measure of human interventions in the natural landscape. Attention to *feng shui* ensures that the spirit of place is not offended by injudicious intervention. The streamline makes such adjustment visible in landscape form, revealing it in the flowing motion of the contour furrow and the curving meander of the regulated and canalized rivers that mark California's Central Valley.

This Modernist vision of streamlined planning reaches an ironic landscape climax where natural streamlines are interrupted, in the concrete monuments of Modernist visionary engineering: mid-century river dams and reservoirs. The most monumental of these landscapes of integrated planning and the synoptic vision is the system of dams and regulated river channels that make up the Tennessee Valley project.[23] Today its makers, the US Corps of Engineers, tend to receive a bad press from America's ecological lobby, which accuses the Corps of arrogant disregard for nature and an assumption of technological supremacy. Whether or not this is true today, the matured landscapes of the Tennessee Valley Authority bespeak a sensitivity in mid-century landscape engineering to natural lines and landforms that corresponded to the broad interest in morphology apparent in mid-century geographic theory, landscape art and photography.[24] The only way that the scope and intention of the TVA may truly be appreciated is to view the landscape from the air or on the map (Fig. 5.4). It is measured in the shapes of the lakes formed behind the river dams, following the natural lines of the land, tracing its contours up the valleys and counterpointing the serpentine forms and natural colours of terrain and water with the linear geometries and concrete tones of the dams themselves.

This Modernist landscape of water regulation reaches its apotheosis in the West, where the dams and power barrages on the Columbia, the Colorado and the Snake systems and California's great aqueducts are the grandest monuments to the concrete sublime.[25] Such a claim is somewhat unfashionable today, and I am not denying that many of the long-term ecological consequences of such radical and large-scale interventions have been seriously detrimental, nor that the sustainability of these landscapes is profoundly uncertain. But it is important to recognize that within the limits of

Figure 5.4 Aerial view of the Hoover Dam under construction, 1936 (Benjamin and Gladys Thomas Air Photo Archives, Spence and Fairchild Collections, Department of Geography, UCLA)

its own understanding of nature, such visionary engineering was intended to harmonize with the natural landscape and to respond to the elemental scale of the Western natural landscapes. Works along the Columbia and Snake rivers, for example, were thought to enhance Western nature and realize its potential through irrigation and electrification: to create a garden in the Pacific Northwest, a utopian landscape worthy of Thomas Jefferson. And the Grand Coulie Dam is indeed appropriate in scale and majesty to the Cascade Mountains, the Columbia Gorge, the forests and deserts that provide its setting.

In the truly arid intermontane West, in the vast continental interior that occupies fully one-quarter of the breadth of the American continent, garden landscape gives way to arid desert. Islands of human occupation, their location authorized by the presence of water, are all that has ever been possible for settlers among 'dry mountain mouths of carious teeth', where scorching sunlight, stinging wind and driving dust repel settled human dwelling.[26] Historically, Native Americans found oases in the cool depths of the canyon walls or replicated their protective shadows in the geometry of adobes set into the mesa. Mormon farmers were among the first to introduce the Europeans' rectilinear geometry into their irrigated plots along the valley washes of Utah. And today pivot irrigation arms inscribe precise circles into land survey rectangles, producing vast mandala markings across the Western landscape. Water, the key to organic life, lies far below the surface in deep aquifers; technology gives it a landscape expression. Here, in the most uncompromising nature that America can offer, we witness an unexpected trace of Leonardo's human microcosm, visible from the air as a circle of green alfalfa set into a square of brown desert.

Eschatology in the measure of landscape

The closest that human technology has come to replicating the energy of the sun is in nuclear fission. Nuclear energy speaks of origins and ends; it has an elemental quality that seems in some ways appropriate to the nature and scale of desert landscapes of the American West. Nuclear testing and development for military purposes introduced a wholly new sense of measurement, and devastated large areas of the intermontane West, littering them with the detritus of abandoned military hardware, unexploded ordnance or radiation, and rendering them dangerous to life for the foreseeable future. At the height of Cold War confidence in the visionary authority of science and technology, atomic hubris led devotees such as Edward Teller to advocate and demonstrate large-scale environmental engineering using atmospheric nuclear explosions.[27] In the same years the same desert landscapes attracted land artists such as Douglas Heizer and Robert Smithson to

100

create vast, precisely measured earthworks that were explicitly works of American landscape, utterly removed from the influences of Eurocentric art movements.

Military presence in the American West is more pervasive than simply its nuclear representation: vast air bases, missile testing grounds and bombing ranges at White Sands and Fort Bliss in New Mexico, Nellis in Nevada, Wendover, Deseret and Dugway in Utah, cover areas comparable in size to some of the smaller European states. The military landscape is not entirely malign; it can conserve ecologies that would otherwise disappear to ranching or recreation. Nor is it always ugly. At Magdalena, stretched under the cloudless skies of New Mexico, the three 15-mile-long tracks of the Very Large Radio Tracking Telescope, each with its nine receiving dishes, record the messages of the planetary bodies, capturing celestial power, and concentrating rays of energy to make sense of the universe. Such activity is a fundamental feature of all cosmologies, archaic and contemporary. Nearby, ancient Hopi and Navajo solar observations are recorded in the locations and forms of their settlements. The cosmographic measure is not lost to the American landscape. We continue to use the same geometries of circles and intersecting lines to take the measure of the cosmos. And the West's desert landscapes remain a magnet for the creative energies released by social, political and above all metaphysical speculation: the great playa of Black Rock Desert in Nevada is the annual location of America's most radical creative arts event, the Burning Man Festival.

California and the West Coast are where the American landscape, read from the perspective of Europe, must end. Beyond is no longer West: it is the 'Far East'.[28] The utopian strain that has counted for so much in the social measure of the American experiment has been particularly intense in the state's mapping and settlement. Indeed, California remained an island on world maps well into the eighteenth century, framed in the classic geographic form of utopia. Spanish missionaries and *Californio* ranchers were never numerous and the measures they inscribed upon the land have been largely erased. It was gold that first attracted Anglos in large numbers to California; climate and visions of self-sufficiency that kept them coming.[29] Southern California was not the only New World landscape to be promoted as America's Mediterranean. Georgia and Texas both enjoyed the same moniker at earlier dates. But California more than any other state promised to fulfil dreams of Arcadian landscape and life, in which a mythic memory of ancient Greece and Rome, or of Renaissance Italy or Spain, could somehow be made real. Traces of that dream abound in contemporary Californian landscapes: in its many and diverse utopian settlements, its housing styles, in the suburban landscaping of the great cities, or in the vineyards of the Napa Valley. But these Mediterranean echoes are found within a distinctively American landscape measure. California's agricultural cornucopia – fields of strawberries or decorative flowers, orchards of avocados, oranges and pistachios, groves of olive trees and vineyards – pays consistent homage to the rigid geometry of the rectangular grid, here defined by lines of concrete irrigation channels or statuesque eucalyptus windbreaks.

More dramatic still are the serried ranks of wind turbines that wholly dominate the San Gorgonio Pass between San Bernadino and Palm Springs. Here, set against the San Jacinto mountains whose dry forms resemble those of Attica or the Peloponnese, on the edge of the nation's most extensive urbanized region, fluted columns rise like the structural elements of ancient Doric temples (Fig. 5.5). Those temples were built at places in the landscape where the *genius loci* of their divinities manifested, and they were minutely oriented to the landscape: to topography, sky and sea. Modern wind farms are equally, if less mythically, sensitive, to the measure of the land. In southern California they convey a sense not only of the Mediterranean dream as one expression of the continuous attempt to make over the New World in the image of an Old World golden age, but of America's authentic cultural achievement in drawing upon the insights of diverse peoples from across the globe and reworking them in a physical landscape that imposes its own measure over all human endeavours.

Figure 5.5 Alex MacLean, Wind farm: San Bernardino County, CA (James Corner and Alex MacLean, *Taking Measures: Across the American Landscape*, New Haven, Yale University Press, 1996)

Conclusion

Measures of time, measures of space; measures of humanity and measures of nature: taking the measure of a geography as vast and varied as America is less a scientific than a creatively imaginative project. From the air a kind of order becomes apparent, above all the insistent autonomy of the land, its capacity to absorb and rework any human imposition giving it meanings much richer than those intended by its human interlopers. The mid-point between earth and stars is a good place to take the human measure of landscape rather than landscapes. We are not tied irrevocably to earth and water; in imagination and vision we can escape to the stars, even if only temporarily. At the same time we cannot subsist in air and fire, but are pulled back insistently to the body of the earth.

In the decade that America was discovered to European eyes, the Italian philosopher Giovanni Pico della Mirandola perfectly articulated this inter-mediate position when he described Adam as a creature of 'indeterminate nature' whom God had placed in the middle of his cosmos:

> Neither a fixed abode nor a form that is thine alone nor any function peculiar to thyself have We given thee, Adam, to the end that, according to thy longing and according to thy judgement thou mayest have what abode, what form, and what functions thou thyself shall desire. The nature of all other beings is limited and constrained within the bounds of law prescribed by Us. Thou, constrained by no limits, in accordance with thine own free will, in whose hand We have placed thee, shall ordain for thyself the limits of thy nature.[30]

During the succeeding half millennium the American landscape has been fundamentally altered, transformed by economic, cultural and techno-logical forces whose origins and scope were only dimly discernible to Pico's world. Today, nature's side of the balance sheet recording this activity is commonly measured as a radical deficit. We are aware, in ways that the Florentine was not, of the limitations on human will imposed by those bonds that unite all nature. Yet his words ring true today as we look down on the American landscape, so powerfully crafted by the social and indi-vidual dreams of generations who have followed him. We can never fully escape our attachments – and our responsibilities – to the landscape, but neither may we stop measuring and dreaming.

6 Wilderness, habitable earth and the nation

> The West of which I speak is but another name for the Wild; and what I
> have been preparing to say is, that in Wildness is the preservation of the
> World.[1]

The last phrase of Henry David Thoreau's proclamation is a *locus classicus* of
American environmentalists' claims for the preservation of geographical
areas and regions of wild and unsettled nature. The words echo like a
mantra through Max Oelschlaeger's philosophical investigation of *The Idea
of Wilderness*.[2] More rarely, if ever, do we consider the first part of Thoreau's
sentence: 'the West of which I speak is but another name for the Wild'.
Thoreau here uses a geographical coordinate (an abstract concept with no
essential environmental reference) as synonymous with 'the Wild', with
wilderness. In so doing, not only does he draw upon what I shall argue is
a characteristically American mode of imaginatively mapping the socio-
environmental geography of nationhood, but he articulates in his own
way key elements of a far broader discourse about chaos and order,
identity and otherness, in which European culture has been engaged
throughout the course of Modernity. It is a discourse which is as much
geographical and cartographic as it is environmental, as much social as it is
natural, and as much about nation and empire as it is about individual and
divinity. I shall argue here that ideas of wilderness are deeply bedded in
the creation of the modern world, but not wilderness alone. The imaginative landscapes of *garden* and *city* are inextricably woven into its
constructions.[3]

I treat Modernity as a European phenomenon progressively apparent in
all aspects of social, economic, political and cultural life from at least the late
fourteenth century, achieving global extension in the course of the twentieth. While recognizing the salience of characteristic elements of Modernism such as capitalist markets, individualism, secularism, applied science

and Cartesian objectivity, as a geographer my focus is on two connected *spatial* dimensions of Modernity. These are the geographical expansion of the world known to and mapped by Europeans, and the extension across that world of those characteristically European spatial forms of collective identity, dwelling and authority: universal empire and nationhood. However we choose to theorize the engine of modern history, it has been the global projection of European power and European modes of life, and the responses to these within and beyond Europe, which have produced the modern concepts of one world and a whole earth within which current environmental debates are located.[4]

Fifteenth-century Europeans who opened the era of overseas expansion and re-stitched what William Crosby has called the severed seams of Pangea held an imaginative world map derived from classical and Islamic Mediterranean science, inflected with the beliefs of late Latin Neoplatonists and early Church Fathers.[5] Variously represented by such writers as Macrobius, Isidore of Spain and later Albert Magnus, and still influential in the mid-fifteenth century, as Fra Mauro's 1459 world map reveals, the world was a single landmass of three continents in the northern hemisphere of a spherical earth (possibly balanced by an antipodean land area below the equatorial Ocean Stream) (see Fig. 1.3). This was the *oikoumene*, the habitable earth. According to the ancient Greek concept of climatic zones, only the central band of this landmass, gathered around the 'middle sea', was truly habitable by civilized people.[6] What existed beyond the *oikoumene* – across the Ocean Sea, in the Antipodes, in the boreal spaces of Scythia and Thule, or in the torrid zones beyond Libya – was unknown. Such spaces could not be the dwelling place of humans, but rather of monsters and marvels whose ancestry dated back to Herodotus, Pliny and Solinus, variously elaborated in medieval bestiaries. The former were the monstrous races: hybrid creatures, neither fully human nor fully animal.[7] Such savage spaces and their inhabitants lay beyond the bounds of civil dwelling, of civilization, whose meaning within Mediterranean discourse lies in the city (*polis*) whose citizens (*cives*) were the cultivators of their native territory. Both classical and biblical traditions placed the city at the peak of a hierarchy of environments at the base of which was wilderness, that is the pre-social, pre-discursive space. The marvellous offered a countervailing theme to the monstrous image of savage nature, suggesting that somewhere at the utmost ends of the *oikoumene* were the spaces of perfection and wonder: the dwelling place of heroes beyond the Pillars of Hercules, the prelapsarian world of the Hyperboreans beyond Thule or the terrestrial paradise of humanity's innocent childhood in the East. Under either interpretation, beyond the habitable lands lay wilderness, which, whether savage or gentle, was a place that lacked pathways and where human dwelling remained impossible, because its lands were not subject to culture and cultivation.

The unknown is necessarily limned in the contours of the known, and thus the geographical discourse of wilderness beyond the *oikoumene* was formulated dialectically out of early modern Europe's own imaginative history and geography.[8] One key source was the memory and experience of empire: the *oikoumene* had once been organized, at least rhetorically, having been proclaimed as a single *civis* under Alexander and later Augustus; and the continuity of their claims was represented by Emperor and Pope within the idea of Christendom.[9] A second was that the two roots of late medieval culture in Europe, classical and biblical, concurred in narrating a progressive history of socio-environmental evolution from chaos to order, from wilderness, through garden to city.[10] United with the memory of empire, the teleological view of history readily generated the belief that the history of civilization follows the course of the sun – towards the West.[11] Renaissance aesthetic theory, drawing upon such readings as Aristotle's *Poetics* and Vitruvius's *Architecture*, mapped homologies between society, gender difference, the human body and the liberal arts into this spatio-temporal sequence. It produced a set of ideal types for the interpretation of ideal landscapes within Europe and in the worlds encountered beyond the *oikoumene*. It is not surprising that cosmographic and cosmological schemes illustrating this discourse were most fully elaborated and imaginatively represented in the sixteenth and seventeenth centuries, the peak period of European geographical expansion.[12]

At the opening of the modern world, then, the European imagination easily constructed a collective cultural identity out of a process of historical narration and geographic mapping. This is apparent in the various poems that wove elements of classical myth and literature, geographical exploration, and patriotism into national epics, such as the Portuguese Luis de Camões's *Os Lusíadas* (1572), the Spaniard Alonso de Ercilla y Zúñiga's *La Araucana* (1569–89), the Italian Ludovico Ariosto's *Orlando Furioso* (1516) and the Englishman Edmund Spenser's *Fairie Queene* (1519). The historical narrative was of imperial order being imposed across the otherness of an intransigent wild nature. This could be an internalized *human* nature – in the original sin, which had to be cultivated out of the child – or *environmental*, in the prehistory of natural wilderness which had to be cultivated in order for civil life and dwelling to be possible. The geographic mapping rendered the city as the locus of human social life, supported and surrounded by cultivated nature and protected from the otherness of wild nature beyond. Wilderness was thus located in the city itself (in its children and lowest social orders), at the periphery of civic space, and at the beginning and end of historical narrative time. It was original chaos and the final act in the cycle of civilization: the war and destruction which urban life and commerce inevitably entailed. The wilderness was thus always correlated with origins and infancy, not only in the sense of innocence, but also of untamed passions and undisciplined conduct, as well as with the end of time and the destruction of the earth. The seeds of social development were located in wilderness; unlike the male gendered city and the feminized garden, wilderness is ungendered space, animal and pre-human.

As their *oikoumene* expanded to embrace an ever-greater part of the surface of the globe, and the configurations of dwelling on what was becoming but the smallest of four (later five) continents changed with its evolving territoriality, so Europeans were forced to adjust their imaginative histories and geographies. This process is well documented, especially for the most radically different of all the geographical discoveries: the literal New World of the Americas.[13] Rather than retelling it, I identify simply aspects of it that relate to wilderness, at two defining moments: the initial discovery of America and the final extension of imperial control over the continental territory of the most successful cultural transplant of European modernity: the United States.

Wilderness and the Caribbean encounter

In the same months of 1492 that Columbus was navigating his first voyage west across the Ocean Sea, one of the most sophisticated projects of European publishing of the incunable period was under way in Nuremburg. Hartmann Schedel's *Chronicle*, published in Latin and German in 1493, actually reports the discoveries (attributing them to Martin Behaim).[14] More significantly for my purposes, Schedel's highly illustrated encyclopaedic universal history opens with a representation of the Christian cosmogony: God's creation of order from the chaos of elements, ending the narrative with a picture of the *oikoumene* drawn from recently published world maps constructed according to Ptolemy's *Geography*. The margins of the map are occupied by dramatic illustrations of the monstrous races, the wild others who existed beyond the margins of dwelling (Fig. 6.1). Among these half-human, half-animal creatures the most terrifying were the anthropophages. Today the word is archaic, having been completely replaced as a designation for eaters of human flesh by the word cannibal. The change is significant, for cannibal derives from the same root as *Carib*.[15] Europe's reading of the New World was from the beginning stretched between the two readings of wilderness we have identified and which can be traced back to the Homeric epics. One posits the existence of an innocent, childlike people inhabiting the islands of the Western Ocean: the Arawaks who live in a golden age of harmony with the natural world. Their mortal enemies were the Caribs, scarcely human, wild warriors and cannibals, bent on conquering, enslaving, and where appropriate *eating* the gentle Arawaks. Such is the enduring power of this narrative that it still informs the historical anthropology of the Caribbean. Disputes over the historical accuracy of this ethnography exist, but are hardly significant for our purposes because, as Hulme notes:

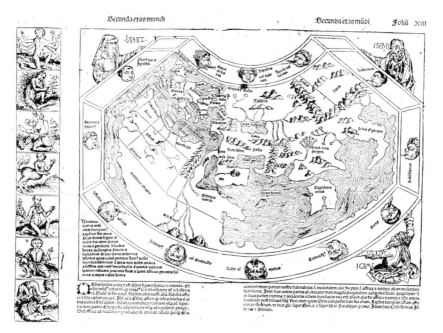

Figure 6.1 Hartmann Schedel, Ptolemaic map of the *oikoumene* showing the monstrous races, from *Liber Chronicarum*, Nuremberg, 1493 (author's copy)

These are ideological, not historical categories and as such have a long history, but the very shock of the contact between Europe and America gave the couplet a new lease of life. These days there are few human societies unaffected by the expansion of Europe and its aftermath, yet ideologically the couplet's demise seems capable of almost infinite postponement. There is always at least the rumour of a last 'primitive' society, inevitably cannibalistic, unvisited by camera or notebook and which, when visited, turns out to have renounced cannibalism only recently. Fortunately, just beyond the next hill is another society, unvisited as yet by anthropologists, and they are still cannibals. Exactly the story Columbus heard nearly 500 years ago.[16]

Cannibalism is of course the ultimate act of the wilderness condition, the very antithesis of civil society, as the fate of the nineteenth-century Franklin and Durren expeditions would later remind 'civilized' Westerners.[17] It is worth remembering, however, that 'cannibals' themselves or a closely related group invariably also turn out, under the different angle of the modern gaze, to be 'gentle peoples of the rainforest', imaginative descendants of the Arawaks, representatives of the 'deep past', living in ecological harmony with their wilderness environment.

In the course of the next three centuries this Manichean reading of trans-atlantic ethnography was extended to the whole of the New World, and into the Pacific. Westward the course of empire took its way, and at every stage the wilderness and its occupants acted as the required counterpoint to a constantly reshaping European identity. As I argued in Chapter 4, attempts to create and sustain an imagined middle landscape, neither savage wilder-ness nor bloated city, run like a yearning refrain: 'back to the garden', through the centuries of New World conquest and incorporation into the European imperium, colonial settlement, and coming to independent nation self-consciousness: from William Penn's plantation, through Jefferson's Land Ordnance Act, utopian plans in Australia, paradise island communities in the Pacific, craftsmen settlements in California, to the contemporary American suburb.[18] But the refrain always competes with war drums pro-claiming the need to conquer the wilderness and its savage inhabitants, while sweet pastoral pipes pick up the elegiac strains of lost innocence and purity in a once-perfect nature and its now-destroyed aboriginal peoples.

Nationhood and wilderness

As Europe's overseas imperium expanded, so the territorial geography of the old continent itself was being internally reshaped into competing nation-states. Wilderness, garden and city are repeatedly mapped into the imagina-tive geography of European nationalism. Romantic nationalism forged myths of harmony between peasant cultures and the natural world.[19] It is a driving force behind the topographic mapping and recreational exploration of European naturescapes from the early nineteenth century until well into the twentieth. Oswald Spengler's reference to 'cultures that grow with ori-ginal vigour out of the lap of a maternal natural landscape, to which each is bound during the course of its existence' derives from the intense German concern to map national time and space. In Britain, the national (and imperial) capital of London was mythically surrounded by the 'home coun-ties' where, as in Kent, the 'garden of England', the nation's heartland is figured as a gentle and domesticated landscape (Fig. 6.2). The kingdom's wild and upland margins were long stereotyped as the haunts of 'ancient Britons', half-savage Celts, Scotch and Irish whose loins contained the seeds of virile nationhood and whose physical prowess rendered them suitable to form the fiercest regiments and colonizing pioneers of empire.[20] Today in these 'marginal' lands are to be found Britain's national parks, areas of upland wilderness to be preserved 'for the nation'.

The designation of land as national park is of course an American inven-tion, subsequently adopted by other nations.[21] Its history and intellectual contexts have been more than amply documented, as has the fact that in the

Figure 6.2 Ordnance Survey, Popular edition map of Nottingham 1:63,360, cover design by Ellis Martin, 1926 edition (author's copy)

USA, and from the late nineteenth century in other regions of large-scale European colonization such as Canada, South Africa and New Zealand, national parks have occupied areas regarded as 'wilderness', the last preserves of lands untouched by the outward expansion of European imperium.[22] In most such areas former modes of dwelling by indigenous peoples had been expunged, often violently, not many years before the declaration of wilderness status.[23] The wild – in the form of both nature and native peoples – was mapped and fenced according to the abstract geometry of the city, as the 1872 Act establishing Yellowstone Park makes clear.[24]

In these same years the radical imposition of modern modes of dwelling across these areas of European settlement found its highest expression in the construction of nations: societies modelled on the form of territorial social organization that had emerged in Europe out of the ruins of medieval and early modern empire. The late nineteenth century, between the unification of Germany in 1870 and the outbreak of continental war in 1914, saw the flood tide of nationalism and colonial imperialism as well as a growing national self-awareness among temperate colonial societies themselves.

One cultural dimension of nationalism was the fascination with autochthonous origins. No longer did nations claim to have been founded by ancient Mediterranean heroes such as Aeneas or Brutus, but were believed to be the authentic expression a 'folk' rooted in the soil of a fatherland. Archaeology, folk revivals, the recovery (more commonly the re-invention) of customs, traditions and costumes: all these were the apparently innocent dimension of a nationalist obsession whose darker side was revealed in the scramble for empire in Africa; in European eyes, this was the last childlike/savage wilderness to be brought into the light of civilization. Among the many ideological pressures which stamped the different strands of European nationalism into a single coin was the belief in social evolution and, in the late nineteenth century, eugenics. The success of Darwinian biology, particularly after Mendelian genetics had provided a clue to the mechanism for the inheritance of acquired traits, gave a huge stimulus to racial theories of cultural difference, which in turn provided an apparently scientific foundation for the differentiation of peoples into distinct nations. Discovering distinct national origins in the distinct natural environments of Europe thus took on a significance scarcely less important than that of differentiating the world's races into the evolutionary hierarchy of civilization.[25]

Wilderness and identity in early twentieth-century America

The United States, since 1776 the very model of a European Enlightenment state, shared this imperative for defining national identity. In the aftermath of the Civil War, the United States not only completed the conquest of a continental empire and the subjugation if not erasure of its indigenous dwellers, but also emerged as an industrial competitor to Great Britain, France or the fast-growing Germany and the destination of a rapidly growing volume of immigrants from Western Europe's peripheries. Grounding national identity within an increasingly diverse and widely scattered community of immigrants, many first-generation arrivals, was a more complex and pressing issue in the USA, however, than in Europe. Certainly, the question of origins could not be settled internally through ethnicity, folklore, archaeology and Dark Age historical myths as was the case in European lands, for America's truly aboriginal peoples were not only seen as non-dwellers, generally believed to be non-cultivators, but were regarded as inferior racial types destined to decline and disappear, and thus physically removed into an apartheid system of reservations. In the resolution of American national origins and identity, as in Europe, natural environment and wilderness were to play a uniquely significant role.

Frederick Jackson Turner's 1893 paper on 'The significance of the frontier in American history,' which dominated American historiography for the following half-century, gave Western wilderness the pivotal role in the building of American national identity. The thesis that in the confrontation between civilization and savagery, on a westward-moving line of settlement, European sophistication was stripped away, to be replaced by a healthy young democracy, drew explicitly on biological metaphors to construct a myth of continuous origin.[26] The final lines of Turner's essay: 'the frontier has gone, and with its going has closed the first period of American history', reflect the fear that the second century of American nationhood could no longer enjoy the benefits of this foundational socio-geographic process.[27] In its more general social expression this fear related less to any geographical frontier itself than to events then taking place in the American city. The link between Turner's thesis and the surge of interest in the preservation of Western wilderness – which reached its peak in the designation of national parks during the presidency of Teddy Roosevelt (1901–9) – has not gone unremarked[28] (see Fig. 4.2). The further link to debates about American racial origins has received rather less comment. Yet it is this connection that allows us to place the American mapping of wilderness more centrally into the long history of European historico-geographic imaginings, and to challenge claims to American exceptionalism.

More immediately significant for their society than the closing of the frontier in the minds of many Americans in the 1890s were the related issues of immigration and nativism. In the years between 1890 and 1914 some 15 million new immigrants entered the USA, cresting at 1.29 million in 1907 alone – a figure not exceeded until the 1980s.[29] In that year Congress set up the Dillingham Commission to investigate immigration. Drawing on the then prevalent belief in 'scientific racism', its report was based upon the belief that since 1880 a 'new' type of person had come to dominate movement into the United States.[30] Unlike the 'old' immigrants from north-western Europe who (it was believed) had arrived for the most part in family groups, entered farming (thus truly 'dwelling' on the land) and 'mingled freely with . . . native Americans [sic]', the new immigrants were unskilled, transient young men, largely from southern and eastern Europe or (in California) from China and Japan, entering urban industrial employment and remaining separate from earlier settled Americans. In but two aspects the report was incorrect in making such a sharp contrast between the two groups. The great wave of immigrants after 1880 entered urban industry, to be sure, but less by choice than because this was now the growth sector of the American economy. But the vast majority (81 per cent) did indeed come from eastern and southern European origins: Austria-Hungary, Romania, Russia, Italy, Greece and Turkey, often removed from their native lands by the very processes of ethnic nationalism under way on the old continent.

The Dillingham Commission was largely a response to the pressures of an American nativism, which grew rapidly from the 1880s, most virulently in the Pacific West where it was directed against Chinese, Japanese, Mexicans and other non-Europeans.[31] In more genteel New England it took the form of Immigration Leagues, led by such Boston intellectuals as the Harvard geologist Nathaniel Shaler, himself an academic conqueror of Western wilderness.[32] Nativists sought to root American national origins in an Anglo-Saxon ethnicity, under which they grouped British, Germans, Scandinavians and, reluctantly, Irish, while excluding Italians, Jews and Slavs as lacking the 'ancestral experience' that characterized established Americans. By 1906 the Immigration Leagues were using explicit geneticist arguments to answer the question of whether America should 'be peopled by British, German and Scandinavian stock, historically free, energetic, progressive, or by slow Latin and Asian races, historically downtrodden, atavistic, and stagnant'.[33] Increasingly their arguments revolved around a belief in racial purity as the foundation of American national greatness, although some did suggest, following F. J. Turner's thesis, that 'racial and hereditary habits' might be overcome by exposure to the American environment.[34]

In such a context the exact coincidence of the great surge of wilderness preservation in the first decade of the new century, so often attributed to the charisma and individualism of such actors as John Muir and Teddy Roosevelt, might perhaps be cast in a broader historical light. The open spaces of the West were as far removed geographically, culturally and experientially from the crowded immigrant cities of New York, Philadelphia, Pittsburgh or Chicago as they could be. From the earliest days their patronage was, necessarily, given the cost of rail fares to reach them, overwhelmingly drawn from 'old' stock and the middle classes, many of whom had been strongly committed to the introduction into the cities of urban parks modelled upon the English picturesque tradition, a style often opposed by newer immigrant citizens who favoured playgrounds and baseball diamonds to lakes, trees and flower beds.[35] Furthermore, the national parks represented, in Turnerian terms, the kind of environment in which earlier – and racially 'purer' – immigrants were believed to have forged American national identity. Western wilderness was not merely the theatre of American empire: it had become the gene-bank of America's national childhood – wild, innocent and free. In its preservation lay the preservation of the nation's youthful loins.[36]

The wild and the child

Youth and innocence are, as I have demonstrated, long-standing features of wilderness ideology. They have been important themes in American cultural discourse from the moment of discovery, but they achieved a special prominence in the *fin de siècle* years of 'new' immigration and national park legislation. Nineteenth-century Romanticism had encouraged a positive reading of childhood as a separate sphere of human experience. Infancy has always been considered close to chaos, thus requiring the civilizing process. In late Victorian bourgeois society, however, like femininity, childhood was regarded as untainted by social artifice and thus embodying a 'moral innocence and emotional spontaneity which seemed increasingly absent from the public realm'.[37] In the closing years of the century the cult of childhood merged on the one hand with that of strenuous exercise as a route to the recovery of unrepressed emotions and the imaginative life, and on the other with ideas of racial purity. As an 1896 reviewer in *The Nation* put it:

> Scientists have informed us that the child alone possesses in their fullness the distinctive features of humanity, that the highest human types as represented in men of genius present a striking approximation to the child type.[38]

In order to sustain the quality of the race in adulthood, eugenically informed educationalists recommended repetition during childhood of the experiences and emotions of primitive ancestors and the cultivation of primitive vitality. Where better to achieve this than in the wildernesses of unspoiled nature, especially for Americans of north European stock, whose ancestral heroes were supposedly pioneers and frontiersmen but who now feared the threat of racial degeneration?

At the century's turn the national heroes of both European imperial nations and the United States were male pioneers and explorers, then conquering the desert outposts of empire: manly men hunting big game in Africa or India, or racing to be first to plant the flag at the ends of the earth.[39] These were the role models for youthful members of such patriotic and character-forming groups as the Boy Scouts and Woodcraft Folk, whose 'Indian knowledge' was an ironic tribute to the peoples who had been destroyed in the process of frontier colonization. The peculiarly American incarnation of this hero type was of course the cowboy, manufactured in the same years through the imaginative art of Frederick Russell and Charles Remington, picked up by Hollywood within a decade, and apotheosized in such white male, all-American heroes as John Wayne.[40]

Without suggesting a conscious policy in the early years of the century to set aside Western national parks as breeding grounds of a particular type of American national character, I suggest that the social and cultural context within which the dominant culture of wilderness in America was established is important in understanding its continuing meanings. Not only does that context suggest a closer link with imperialist, xenophobic and racist features of American nationalism than many Americans would feel comfortable with today, but it locates the discourse of wild nature in the early national park movement within a historically deep-rooted Western cultural tradition spanning ideas of order, civilization and empire, and with a racialized and gendered national identity. Contemporary appeals to the idea of wilderness are unlikely to be free of such visceral attachments, as the vehemence with which wilderness debates are conducted within the United States today perhaps indicates.[41]

Wilderness ideas today

Today, designated wilderness preserves in the United States occupy some 80 million acres, or 3.5 per cent of the national territory, largely the result of sustained pressure upon the federal government since the 1930s by groups such as the Sierra Club, the National Audubon Society and the Wilderness Society, pressure which achieved its greatest success in the 1964 Wilderness Act. Membership of these groups has been closely related in personnel and intellectual pedigree to nineteenth-century nature advocates pressing for urban, and later national, parks to improve the material and spiritual health of urban America. A sustained, but empirically unsubstantiated, belief in the spiritual power of wilderness as an antidote to 'the terrible divorcement from a native environment' occasioned by urban living remained central to their arguments for wilderness at least until the past decade, and continues even today in some quarters.[42] Actual use patterns hardly support the claim that such wild areas offer an alternative for disadvantaged urbanites and minorities to 'hospitals, mental institutions, clinics, prisons, and cemeteries'.[43]

A striking sociological feature of any visit to an American national park or forest, and even more to a designated wilderness area, is how radically unrepresentative of the vibrant social and ethnic mix found in America's cities are their visitors. Young, white, fit and healthy, if not wealthy, middle-class families, both American and increasingly also international tourists, overwhelmingly dominate the camping sites and visitor centres, while the hikers and backpackers on the wilderness trails are almost invariably youthful members of this same group, whom the Immigration Leagues of the early century had proclaimed to be original Americans. Rising charges in national and state parks have further restricted access by income. The median age of national parks and wilderness area visitors has been slowly rising as the baby-boom generation passes through the American population. This bulge group came of age in the years surrounding passage of the Wilderness Act and participated in the first wave of ecological protest during the late 1960s. Highly elaborated codes of dress and behaviour reflect the quasi-military origins of the parks and restrict communal activity in favour of more individual recreational pursuits.[44] Wilderness literature also suggests that many features of the American cult of nature have remained little changed over the course of more than a century. Popular wilderness literature celebrates a particular form of youthful, essentially individualistic, solitary and male physicality, a muscular stoicism often combined with a conceit of childhood innocence and mystical introspection in the wild which privileges visceral, primitive vitality over intellectual and emotional urban lassitude in our experience of the world.[45]

At the more scholarly level, wilderness writing no longer stresses the direct effects of nature experience on individual and social health, although its beneficial impact upon personal psychology is still frequently proclaimed. The argument for wilderness has shifted in some degree towards a progressive rhetoric borrowed from civil rights and liberation movements, extended beyond human groups to an inclusive nature granted its own rights. This is justified on the grounds that evolutionary theory has undermined any philosophical or theological arguments for human exceptionalism within the created order.[46] Wilderness thinking thus remains profoundly coloured by a commitment to natural order, and by a desire to read the social into a narrative of the natural, or, in its own terms, a unitary narrative of both.[47] The argument also remains dominated by the uncritical acceptance of a bio-ecological paradigm, whose own intellectual roots are tangled up with much of the unsavoury racial and eugenic theorizing of the early twentieth century.[48] In a philosophically informed historical study of the 'wilderness idea' Max Oelschlaeger, for example, whose work is in most respects profoundly sceptical of received wisdom and which self-consciously proclaims the post-modern collapse of scientific foundationalism, treats evolutionary theory as foundational, and the discourse of ecosystems and entropy as unproblematic representations of nature rather than as currently accepted metaphors whose half-life is unlikely to be longer than the mechanical metaphor he so properly condemns. His narrative, moving from the 'deep time' of the Palaeolithic to the forecast future of global ecocatastrophe, via the Neolithic garden and Modernist city, follows the well-worn trajectory of the West's mythical narrative of social evolution.[49]

Perhaps the most disturbing aspect of much conventional wilderness ideology is its unreflective imperialism. I have sought to emphasize the specifically American construction of the modern wilderness idea, while placing that construction within a long history of European imaginative mapping and dwelling in the world. Since the early 1970s one of the most popular icons of the wilderness movement has been the image of the earth from space, which has replaced conventional cartographic representations of the earth[50] (see Fig. 1.4). The environmental interpretation of the Apollo earth images of the global *oikoumene* as a vulnerable earth, isolated in a secular void of dead space, draws upon specifically American yearnings for socio-environmental perfection, isolation and the need for moral regeneration through recognition of nature's claims. Above all, such an interpretation of 'Whole-Earth' readily serves to justify what are effectively imperialist interventions anywhere across the globe in favour of environmental and wilderness preservation. It is ironic that at the historical moment when the system of territorial states and national integrity pioneered by Enlightenment beliefs in human rights, social perfectibility and the creation of the liberal state has finally been extended across the globe through the process of decolonization and the destruction of the old European *imperium*, it should be undermined by the ideology of an empire of nature. The maintenance of environmental order in this empire is policed by spokespeople and activists drawn principally from the old imperial nations, often using similar language of pioneering adventure, in the same geographical theatres of snow, desert and jungle as their colonial forebears.

Conclusion

The Wild and the West were intertwined as imaginative categories within the European imagination for centuries before Thoreau, but less as a place of human dwelling than for mapping visions of social and environmental order, and as a context for personal development and group identity. The theatre of wilderness remains as much a representation of contemporary social and psychological formations as a bedrock reality. It should therefore neither surprise, nor perhaps unduly worry us, that in a global political economy with its consumerist creation of meaning, the commodification of wilderness parallels that of the city, and that the fantasies traditionally associated with the wilderness idea produce such phenomena as urbanization in Yosemite, international 'cannibal tours', and Mount Everest's waste disposal problem.[51] Historical understanding suggests caution in response to the current wave of anti-immigration sentiment in the United States, some of whose strongest voices come again from the West and include prominent members of the Sierra Club.[52]

In the final years of the twentieth century and the opening years of the new millennium the debate over wilderness has indeed opened up, as I have suggested, initiated by a growing recognition among environmental historians of the social and historical frames that encompass American nature. The powerful role of visual images in making the influential Sierra Club vision of wilderness has been dissected by writers such as Finis Dunaway, while Byron Wolfe and his collaborators' rephotographing projects of iconic Western landscapes graphically expose the continuity and inevitability of anthropogenic environmental change in landscapes supposedly preserved from human presence.[53] The wilderness activist Rebecca Solnit has suggested that visitors to wilderness in America are becoming more diverse and their responses less narrowly determined by the conventional wilderness ethic.[54] Yet the vehemence of response to these alternative visions of American wilderness landscape denotes the authority that the discourse I have outlined here still commands in American popular culture.

IV John Ruskin: vision, landscape and mapping

7 The morphological eye

Every irregularity on the surface of the earth, every hummock and valley, stream and pond has a cause, perhaps discoverable. Also, every such irregularity has an effect, perhaps traceable. And some people are interested in the causes of the hummocks, others in their effects. When you tell the hummocky-cause people (geomorphologists) that there is another way of looking at things they say they know all about that, but this is the way to begin . . . Once a week we leave the School of Geography and proceed (carrying lunch) on an excursion – ostensibly they are all for the study of geomorphology – of land-forms and their causes. We have been over most of the country now within a radius of 20 miles or so. And week by week the joy of the excursions has increased and the grip of geomorphology has lessened. Beauty has deepened – blue of hyacinths – delicate tracery of young beech leaves against the sky, the burning bush of the gorse, sky larks and cuckoos, wind on the open downs, mingled rain and sunshine on old stone, thatched roofs and cottage windows, rainbows and the glory of golden sunsets have met us week by week & called with a more insistent voice that we look at them & heed them . . .[1]

These lines, written by a student on the 1914 Oxford Vacation Course in Geography, echo sentiments that the nineteenth-century art critic and social commentator John Ruskin would have recognized, and of which he would have approved. Ruskin himself considered and wrote about the educational value of geography. His years as Slade Professor at Oxford, during which he sought to 'establish both a practical and critical school of fine art', leaving to the university a collection of images and artefacts for art education, overlapped with the establishment of geographical teaching in the university. Ruskin died in 1900, the year that the first cohort of Oxford students received their diplomas in the newly established discipline of geography.

The examination they sat for the School of Geography at Oxford included questions about the influence of the Alps and Rhine on the course of human history, and the implications of geological faults for the development of surface landforms. Reading that exam paper, a student who had attended the lectures on landscape or drawing that Ruskin had given at Oxford in the 1870s or 1880s might have recognized in the 'new' geography some themes from those earlier lectures.

Ruskin argued passionately 'that the sight is a more important thing than the drawing', and to 'teach drawing that my pupils may learn to love Nature, [rather] than to teach the looking at Nature that they may learn to draw'[2] (Fig. 7.1). In this chapter I explore his connections with geography. These include both the formal discipline then being shaped in British universities, and the 'geographical imagination': that urge to explore and describe a mutual shaping of land and life into landscape. Heirs to Romantic nationalism and the cult of the picturesque, powerfully influenced by physiographic principles, and committed to the intellectual and social issues of their day, both John Ruskin and Oxford's 'new' geographers had much in common. Juxtaposing their work generates insights into how the study of landscape and maps in geography offered a bridge between art and science, and a focus for social thought and criticism. There are few direct connections between Ruskin and geography, but the two are related: locally, through common involvement in an intense debate about art and science in later Victorian Oxford; and more generally, through an attachment to maps and landscape as ways of seeing, thinking and learning about nature and society.

Figure 7.1 John Ruskin, Watercolour sketch of the Alps (Teaching Collection, Ashmolean Museum, Oxford)

Physiography

Before examining the details of art and geography teaching in *fin de siècle* Oxford we should note an immediate pedagogical context. This was the widespread popularity and influence of Thomas Henry Huxley's 1869 writing on *Physiography*, originally delivered as extension lectures to working men and published in book form in 1877.[3] Both John Ruskin and Halford Mackinder, Oxford's first Reader in Geography, also participated in the Victorian educators' commitment to working men's learning. For Huxley physiography was a practical method for teaching the causes and interconnections of physical phenomena, which constructed scientific knowledge from the students' direct experience of the world rather than from abstract theories. Its 'fundamental principle was to begin with observational science, facts collected; to proceed to classificatory science, facts arranged; and to end with inductive science, facts reasoned upon and laws deduced'.[4] The book's impact was very considerable: 'within six months its circulation had become "enormous" '.[5]

John Ruskin was appointed to the Slade Professorship of Art at Oxford in 1874, only four years after Huxley's publication appeared and at the height of its popularity. By this time Ruskin himself had devoted nearly 40 years of writing and lecturing to the proposition that true landscape art was as rigorous as science in its close observation of natural phenomena, and was charged with faithfully representing law-like qualities of a nature that reflected divine purpose. I discuss this 'mythopoeic science' in more detail below. At Oxford Ruskin had the opportunity to establish a rigorous curriculum to instruct students in these artistic principles. If he did not draw directly on Huxley's ideas, both his approach to nature and his pedagogical vision had much in common with them, although Ruskin never embraced the Darwinism that was central to Huxley's thinking. The first geographical teachers at Oxford were much more directly influenced by Huxley's work. Halford Mackinder regarded physiography as integral to what he conceived as 'the new geography', although he would later seek to distance his discipline from physiography's exclusively physical concerns. Andrew John Herbertson, whom Mackinder appointed to teach the Oxford curriculum, had worked closely with Huxley's student, the architect and visionary planner Patrick Geddes, for whom regional survey was a critical physiographic technique. Herbertson established field observation as a foundation for geographical training in the Oxford curriculum.

Huxley had conceived physiography as the foundation for a scientific education that led on to understanding the laws of physics, chemistry and biology. The geographers, concerned to distinguish their field from geology and to position their discipline as a unifying study of the physical environment and human society, took a more synthetic approach. As the anonymous writer in 'Geophil' quoted above indicates, this brought the experience of geography very close to what Ruskin sought to imbue in his art students.

Graduates of Christ Church: art, science and religion in Victorian Oxford

John Ruskin first went to Oxford as an undergraduate at Christ Church. His earliest published essays were written there in 1837 and 1838 and collected as *The Poetry of Architecture: the architecture of the nations of Europe considered in association with natural scenery and national character*. Their aim was 'to trace in the distinctive characters of the architecture of nations, not only in its adaptation to the situation and climate in which it has arisen, but its strong similarity to, and connection with, the prevailing turn of mind by which the nation who first employed it is distinguished'.[6] The youthful John Ruskin embraced the Romantic-Nationalist belief that each nation has a singular culture, apparent not only in its language but in the handiwork of ordinary folk, rooted in their native soil: 'man, the peasant, is a being of more marked national character, than man, the educated and refined'.[7] The visible landscape, shaped over centuries by peasant hands working with local materials, is pleasing to the eye – picturesque – because it expresses a harmony of nature and culture. These arguments were elaborated in the early volumes of *Modern Painters*, the defence of Turner's landscape art, initially signed by a 'Graduate of Oxford', that established Ruskin's reputation. Similar arguments under-wrote the work of nineteenth-century German geographers as they mapped the field patterns, farmhouses, barns, fences and hedges of settlement types in an attempt to plot the distribution of the German *Volk*, or of French geographers seeking to delimit the characteristic *pays* or local landscapes of rural northern France.[8]

Geography's progress in the German universities was one of the factors leading to the Royal Geographical Society's support in 1887 of a readership in geography at Oxford University. Halford John Mackinder, the 26-year-old appointee, was also a Christ Church graduate. Mackinder's student years had coincided with the then 64-year-old Ruskin's second period as Slade Professor and his 1883 lectures on *The Art of England* and *The Pleasures of England*. These were delivered at the Oxford Museum, that architectural monument to the synthesis of art, science and religious belief at the heart of natural theology, on whose design Ruskin and Christ Church's Henry Acland had collaborated in the 1850s. There is no record of Mackinder's having attended Ruskin's talks, but in the museum's laboratories the young Mackinder studied physical and biological sciences and specialized in animal morphology. Ruskin's opposition to the proposed physiology laboratory at the museum because of its connections with Darwinism and especially vivisection, voiced in those lectures, became one of the grounds for his resignation from Oxford in 1885. By then, Mackinder had graduated and was teaching geography to working men at the Rotherham Mechanics' Institute and framing the argument he would put to the Royal Geographical Society in 1886 for a 'new geography': 'one bridge over an abyss [between the natural sciences and the study of humanity] which in the opinion of many is upsetting the equilibrium of culture'.[9] It was a powerful perform-ance, enhanced by Mackinder's graphic morphological account of the chalk landscapes of south-east England.

Among those who shaped the case for geography at Oxford was J. F. Heyes FRGS, a tutor in 'experimental science' at Magdalen, from whose 'plea for geography' in the December 1886 edition of the Oxford Magazine Mackinder took both the metaphor of geography as a bridge between the arts and sciences and the idea of the 'philosophical geographer'.[10] Heyes was an ordained pastor, moved by Andrew Mearns's 1883 work, The Bitter Cry of Outcast London, to leave Magdalen to work with the Oxford Settlement in Bolton. He continued into the 1920s to promote 'Geosophy: the ideal and more philosophical side of geography', concerned to establish harmonious relations between land, work and folk.[11] This socially and morally relevant geography, akin to the ideals promoted by Patrick Geddes and the LePlay Society, was to be taught through morphological studies that connected the forms and distributions of physical geography to those of human occu-pancy. More directly associated with evolutionary theory and in practice with imperial education, Mackinder's 'new' geography also sustained a deep commitment to regional study, while less socially committed to the disadvantaged than Heyes's vision. As the School of Geography's historian, Ian Scargill, points out: 'the Oxford School has always emphasised a need to study the interrelationships of the physical and human aspects of the subject and to observe this interaction within particular areas of the earth's sur-face.'[12] Ruskin, like Huxley, had a direct influence on the development of Geddes's scientific and educational ideas.

Mythopoeic science and the art of landscape

Francis O'Gorman has claimed that Ruskin's 'prickly, provocative and mythopoeic science' had a much greater impact on later Victorian education than is generally acknowledged.[13] The fundamentals of that sci-ence – observation and morphology – united the ruling passions of Ruskin's intellectual life: Alpine geology and J. M. W. Turner's landscape painting. Their parallels with physiographic method are obvious. Ruskin claimed in Deucalion, the text of his 1874 Slade lectures, that 'the position which it was always the summit of my earthly ambition to attain [was] that of President of the Geological Society'.[14] The claim should be set against that made in his *Lectures on Landscape* three years previously:

> the interest in landscape consists wholly in its relations either to figures present – or to figures past – or to human powers conceived. The most splendid drawing of the chain of the Alps, irrespective of their relations to humanity, is no more a true landscape than a painting of this bit of stone.[15]

Turner alone, he believed, had truly penetrated the meaning of the mountains: 'drawings of the Alps by Turner are in landscape, what the Elgin marbles or the Torso are in sculpture'.[16] What Ruskin admired and constantly emphasized in Turner's landscape art was its truth, not in the sense of an analytic and mimetic rendering of physical facts in the visible world, but of a synthetic grasp of the physical form of the world as a moral imperative: designed for human life and inhabited by men and women.

> Landscape painting is the thoughtful and passionate representation of the physical conditions appointed for human existence. It imitates the aspects, and records the phenomena, of the visible things which are dangerous or beneficial to men, and displays the human method of dealing with those, and of enjoying them or suffering from them, which are either exemplary or deserving of sympathetic contemplation.[17]

Initially grounded in an evangelical natural theology, by the time of his Oxford lectures Ruskin's evangelical doctrine had been chipped away by the geologists' hammers and, in the face of secular Darwinism, had broadened to embrace a more general mythic truth:

> Proserpine and Deucalion are at least as true as Eve or Noah; and all four together incomparably truer than the Darwinian theory. And in general, the reader may take it for a first principle, both in science and literature, that the feeblest myth is better than the strongest theory: the one recording a natural impression on the imaginations of great men, and of unpretending multitudes; the other, an unnatural exertion of the wits of little men, and half-wits of impertinent multitudes.[18]

Lifelong, daily Bible study and a wide literary debt to authors read in their original languages – Homer and Hesiod, Dante and Shakespeare among them – provided the mythical resources for Ruskin's science. But his landscape study was rooted in field observation, in visual practice. An eye formed by early training in the theory and practice of picturesque art and a fascination with geological and meteorological observation were refined by patient, detailed sketching and note taking. Field study was the principal purpose of Ruskin's almost annual continental tours. Their itineraries scarcely varied: through eastern France and the Jura, the Rhineland and the cities of upper Italy, staying above all in Alpine Switzerland, at the Jungfrau, Fribourg and Chamonix. Drawing on his field observations, Ruskin engaged continuously in discussion with scientists as well as critics. On mountain form and glaciation he learned from, but never feared to criticize, the great Alpine geologists, Agassiz and de Saussure who, Ruskin claimed, 'had gone to the Alps, as I desired to go myself, only to *look* at them, and describe them as they were, loving them heartily – loving them, the positive Alps, more than himself, or than science, or than any theories of science'.[19] He wrote extensively on geological uplift and erosion and examined the effects of ice movement and denudation, presenting his work at the Royal Institution and the Geological Society where he enjoyed the support of his friend Sir Roderick Murchison, the 'dominant and dominating figure in the [Royal Geographical] Society at mid-century'.[20]

Behind both nineteenth-century geography and John Ruskin's understanding of the physical earth, among many other influences, lay the learning and the spirit of Alexander von Humboldt. Ruskin shared the German's commitment to travel, to measurement – especially of mountains and meteorology – to synthesis, and to the utility of science to human concerns, so much so that a recent writer has seen parts of Ruskin's *Modern Painters IV* as a contribution to international Humboldtian science.[21] Strongly influenced by Humboldt's *Personal Narrative* which he had read before entering Christ Church, Ruskin would reject the later *Cosmos*, which he encountered in translation about 1846, because of its vision of a world 'animated by internal forces' rather than by divine will.[22] In his Oxford lectures, however, Ruskin included von Humboldt among 'the multitude of quiet workers on whose secure foundation the fantastic expatiations of modern science depend for whatever good or stability there is in them'.[23] This caution may account for Ruskin's failure to engage with von Humboldt's own detailed history of landscape art in volume II of *Cosmos* which I discussed in Chapter 2. There, von Humboldt argues along lines that anticipate Ruskin's own project in *Modern Painters*, that artistic appreciation of natural scenery is a fundamental stimulus to imagination, exploration and scientific knowledge of the physical universe.

Globe and local landscape in Ruskin's geographical imagination

At first sight, Ruskin's restricted European itinerary, with its echoes of the aristocratic Grand Tour, and his minute attention to the morphology of individual rocks, plants, twigs or birds' wings, seem parochial and narrow when compared to von Humboldt's odyssey through Mexico and South

America, and to the conceptual sweep of *Cosmos*. But Ruskin's geographical imagination quickly transcended picturesque localism to reach for more universal claims. To be sure, the Englishman never visited the Americas, and he famously compared humanized European landscape to the gloom of American forest wilderness (also a Humboldtian trope), finding the latter wanting as a subject for landscape because of what Ruskin took to be its historic lack of human association. 'Niagara, or the North Pole and the Aurora Borealis, won't make a landscape; but a ditch at Iffley will, if you have humanity enough in you to interpret the feelings of hedgers, ditchers and frogs', he told his Oxford students.[24] If ruthlessly Eurocentric, Ruskin's vision was also always globalizing. Already in *The Poetry of Architecture* we find him generalizing landscape types according to picturesque criteria of climate, topography and dominant colour tones: the 'woody or green country' dominated by woodland and pastoral economy under a temperate climate, the 'cultivated or blue country' of rich champaign lands and low hills dominated by arable cultivation, the 'wild or grey country' of unenclosed, treeless and plateaued ploughland, and the 'hill or brown country', little populated mountain regions of varied and accidented topography.[25] Such classifications recur throughout Ruskin's writings: Venice is figured as the centre of the world, the meeting place of global cultural streams: Latin, Greek and Arab. And the most evocative passage in 'The nature of Gothic' describes an imaginative flight over 'the zoned iris of the earth', where Ruskin invokes a poetic geography from 'the Mediterranean lying beneath us like an irregular lake, and all its ancient promontories sleeping in the sun', over 'the vast belt of rainy green, where the pastures of Switzerland, and poplar valleys of France, and dark forests of the Danube and Carpathians stretch from the mouths of the Loire to those of the Volga', and 'farther north still, to see the earth heave into mighty masses of leaden rock and heathy moor . . . and splintering into irregular and grisly islands amidst the northern seas', until 'the wall of ice, durable like iron, sets, deathlike, its white teeth against us out of the polar twilight.' This physical description, which reworks the zonal theory of classical geography, is succeeded by an equally evocative geographical sketch of the distribution of vegetable, animal and human life as they respond in their different but parallel ways to 'the statutes of the lands that gave [them] birth'.[26]

Ruskin planned but never executed a historical geography of Switzerland. The closest he came to a 'geographical' text was *The Bible of Amiens*, written and published in 1880, between his two periods as Slade Professor. The work fulfils the description of geography's task laid out six years later by Heyes: to provide 'the physical basis of history', in Ruskin's case the history of Franco-German relations as a foundation of European culture. Typically, Ruskin opens his study with a powerful evocation of place. He invites his English reader to leave the London–Paris train at Amiens and stand before its

Figure 7.2 Amiens cathedral

vast and beautiful cathedral (Fig. 7.2). He asks how this work of art should come to have been built here, in 'the central square of a city which was once the Venice of France'. His answer, woven through an iconographic analysis of the monumental sculptures of Amiens cathedral, is a cultural geography of Europe from its late Roman origins to the recent defeat of France at Sedan:

> The Life, and Gospel, and Power, of it, are all written in the mighty works of its true believers: in Normandy and Sicily, on river islets of France and in the river glens of England, on the rocks of Orvieto, and by the sands of Arno. But of all, the simplest, completest, and most authoritative in its lessons to the active mind of North Europe, is this on the foundation stones of Amiens.[27]

Ruskin, maps and the Oxford School of Geography

Ruskin's argument in *The Bible of Amiens* is grounded in an interpretation of France's physical geography, which he illustrates in a series of thematic maps showing the rivers of France in relation to the country's historical territorial divisions (see Fig. 8.1). These in turn are related also to the geological structure of the country. In a *Fors Clavigera* letter of October 1884,

Ruskin discusses these maps and geographical education more generally. Noting that the subject is well taught in elementary schools, he is scathing about geography at more advanced levels. A major factor, he opines, is the inadequacy of atlas maps. In preparing the maps of France he had turned to *The Harrow Atlas of Modern Geography* (1856) and is critical of both its conception and design. The map is dominated by railway lines and manages to indicate Mont Blanc in two separate locations:

> neither of the Mont Blancs, each represented as a circular pimple, is engraved with anything like the force and shade of the Argonne Hills about Bar-le-Duc; while the southern chain of the Hills of Burgundy is similarly represented as greatly more elevated than the Jura. Neither the Rhine, Loire, nor Seine is visible except with a lens; nor is any boundary of province to be followed by the eye; patches of feeble yellow and pale brown, dirty pink and grey, and uncertain green, melt into each other helplessly across wrigglings of infinitesimal dots.[28]

Ruskin pleads for 'proper physical maps' and 'proper historical maps', calling on the Geological Society to produce 'true models to scale of all the known countries of the world':

> These, photographed in good side light, would give all that was necessary of the proportion and distribution of the mountain ranges; and these photographs should afterwards be made the basis for beautiful engravings, giving the character of every district completely, whether arable, wooded, rocky, moor, sand, or snow, with the carefullest and clearest tracing of the sources and descent of its rivers; and, in equally careful distinction of magnitude, as stars on the celestial globe, the capitals and great provincial towns, but also absolutely without names or inscriptions of any kind.[29]

Such maps and models, which anticipate the topographic plaster casts that illustrated physiographic principles and became common in university geography departments in the early twentieth century, should, Ruskin claims, be free from text, for the student should know already the location of places. For Ruskin, the educational value of maps lies in their illustrating graphically broader spatial relations, and is as much artistic as geographical. 'Once a system of drawing rightly made universal, the hand-colouring of these maps would be one of the drawing exercises, absolutely costless, and entirely instructive.'[30]

Henry Acland's promotion of Ruskin's professorship at Oxford was part of the former's plan for educational reform. And while Ruskin was appointed to lecture on art, he rapidly conceived a reformed art education along the lines he had pioneered in his work at the London Working Men's College and in the Guild of St George in the 1850s and that he summarized in *The Elements of Drawing* (1857) and *The Laws of Fésole* (1875). In his 1871 'Lectures on Landscape' he summarized the goal of landscape art, which would be the principal object of his art teaching:

> The thoughtful and passionate representation of the physical conditions appointed for human existence. It imitates the aspects, and records the phenomena, of the visible things which are dangerous and beneficial to men, and displays the human method of dealing with these, and of enjoying them or suffering from them, which are either exemplary or deserving of sympathetic contemplation.[31]

Five sets of drawings, prints and photographs used in his teaching are preserved at the Ashmolean Museum in Oxford.[32] They comprise overwhelmingly either pieces of Ruskin's own work or pieces he had commissioned. Items include his wonderfully detailed study of gneiss rock at Glenfinlas, and sketches and watercolours illustrating cloud types, water, wild flowers and Alpine glacial features. Others are architectural or show cultural landscapes: for example, Fribourg walls, which have 'flexible spines, and creep up and down the precipices more in the manner of cats than walls', Verona, the Roman Campagna or Chamonix[33] (see Plate 5). The five series were constructed to support a comprehensive educational programme in which draughtsmanship and art teaching always had 'the ulterior object of fixing in the student's mind some piece of accurate knowledge, either in geology, botany, or the natural history of animals'.[34]

The collections do not include maps, but Ruskin took a very direct interest in the pedagogical value of cartography. He devotes a whole chapter of *The Laws of Fésole* to map drawing as such, here too criticizing inaccuracies in existing maps and atlases. He develops a typically idiosyncratic system of global latitude and longitude lines based on a prime meridian running through Fésole ('Galileo's line'), to be engraved on a globe:

> And, on this globe I want the map of the world engraved in firm and simple outline, with the principal mountain chains, but no rivers, and no names of any country; and this nameless chart of the world is to be coloured, within the Arctic circles, the sea pale sapphire, and the land white; in the temperate zones, the sea full lucia, and the land pale emerald; and between the tropics, the sea full violet, and the land pale clarissa.[35]

The globe would permit the student to produce azimuthal maps at different scales centred on any chosen point on the earth's surface. For these Ruskin develops an entire geographical nomenclature, reflecting his conception of the historic and mythic connections between physical geography and human life, and woven into exercises in spherical geometry. His proposals reflect a fascination with geometrical problems that went back to his days at Christ Church.

Like Ruskin's, Halford Mackinder's Oxford appointment was to give lectures. But Mackinder, too, was an educationalist who had developed his ideas through instructing working people, and he also based his teaching ideas on that experience. Within little more than a decade Mackinder and Andrew John Herbertson had collected the materials and designed the curriculum for a Diploma in Geography, and for the teachers' vacation Summer Schools begun in 1904. In 1899, when the course of study for the one-year Diploma was fixed, it comprised:

> The principle and chief facts of Geomorphology. Rivers and river basins. The Coastal belt. Mountains. Areas of elevation and depression.
>
> Surveying, as practised by explorers . . .
>
> The geographical coordinates. Map projections on the plane, the cylinder and the cone . . .
>
> The determination of position by astronomical methods. Simple observations for latitude, longitude and true bearing. Methods of depicting land relief . . .
>
> The reduction and generalisation of maps . . .
>
> The distribution of solar energy on the rotating earth and the results of the circulation of air and water . . .
>
> Meteorological and hypsometrical observations
>
> The climatic provinces of the earth. Compilation and use of charts of weather and climate.
>
> The physical conditions of the Oceanic abyss . . .
>
> The chief generalisations regarding the distribution of animals and plants according to species and associations.
>
> The chief facts of Anthropogeography. The geographical distribution of men according to number and race. The influence of physical features in determining the position of settlements and roads.
>
> Outlines of the historical geography of Europe and the Mediterranean landscape, considered in relation to physical features . . .
>
> The history of geographical ideas. Outlines of the history of Discovery. The distribution of place-names in the Old Continent according to origin and meaning.[36]

Except for the absence of sketching, draughtsmanship and colour, it is hard to imagine a curriculum more attuned to the principle interests displayed in Ruskin's writings and teachings. Indeed the examination paper of 1900, the year of Ruskin's death, invites the student to 'Discuss the influence either of the Alps or of the Rhine on the course of human history in Europe', a perfect invitation for a Ruskin essay.

As outlined, the diploma programme is theoretical and classroom-based with little reference to field study. But field observation was definitely included: 'a district near Oxford will be practically surveyed with a view to its complete geographical description by each student'. And we know from the vacation programme that field-sketching was a significant element in geographical training:

> Certain days, however, will be given wholly to surveying and drawing sketch-maps in the field, and on one day an excursion will be arranged to places of geographical interest. In addition there will be several shorter afternoon excursions to places in the immediate vicinity of Oxford.[37]

The foundation of Oxford geographical education, as the order of both the Diploma programme and the student account quoted at the start of this chapter make clear, was geo-morphology, the study of landforms and their connections with human culture and history. Morphology is the theoretical link between Ruskin's drawing classes and Mackinder's geography; field observation, survey and mapping are the methodological connections.

It is the weekly field excursions that are described in 'Geophil'. That account reminds us that geographical field practice, as John Ruskin always recognized, could lead – at least for some students – beyond the close observation of form and the search for causal connections to an appreciation of what Ruskin regarded as the morality of landscape: the beauty and truth of human existence in nature.

> It is needful evidently that we should have truthful maps of the world to begin with, and truthful maps of our own hearts to end with; neither of these maps being easily drawn at any time, and perhaps least of all now – when the use of a map is chiefly to exhibit hotels and railroads; and humanity is held the disagreeable and meanest of the Seven Mortal Sins.[38]

Conclusion

Both Ruskin and the Oxford geographers saw their work in the university as educational. Ruskin's art instruction was not aimed at developing personal expression or individual creativity but at developing his students' vision of the natural world and the place that human creatures occupy within it. The geographers were not concerned to establish a research discipline aimed at scientific discovery, but as their physiographic, survey and field methods demonstrate, they wanted to produce perceptive, informed and reflective citizens from the undergraduates in their care. For both parties training of vision was important: a training that involved field observation and the recognition of form and morphological relations, in order to generate questions of cause and effects that went beyond natural science to human and social concerns.

The type of instruction that the founders of Oxford's School of Geography initiated continued to determine undergraduate geographical experience in the School well into the post-war years. My own geographical education there in the mid-1960s was still strongly influenced by the Oxford School's founding vision (although the syllabus would be radically reformed soon thereafter). John Ruskin did not figure in the teaching, but it was perhaps the ideals he shared with the Oxford geographers that brought me to his writings as a graduate student. While education and practice in both art and geography have changed radically since the closing years of the nineteenth century, as have the worlds they both study, the common vision articulated by John Ruskin and the early Oxford geographers may still have something to tell us today about geography as a way of seeing and engaging with the world.

8 Ruskin's European visions

John Ruskin's approach to landscape, as I have argued in the previous chapter, can be compared closely to the late Victorian educational vision of geography promoted by Halford Mackinder and Andrew John Herbertson. Like the Oxford School's first Diploma syllabus, it was at once global and local, European and British in focus. Ruskin's consistent purpose was to render Britain and its people fit to inherit what he regarded as the spiritual and cultural mantle of a European and Christian civilization. For Ruskin this was best understood as the outcome of divinely guided geographical and historical forces. Ruskin himself spent much of his life travelling through landscapes and living in cities long mapped out by the Grand Tour that supposedly completed the education of Britain's ruling class, and he read both ancient and contemporary European languages with ease. The influence of his writings on continental Europe was by no means negligible: his ideas on architectural preservation were widely credited with altering continental conservation practices, and his impacts on the thought and work of such literary giants as Tolstoy and Proust are well recorded. Yet Ruskin today both reads and is read as a uniquely English (at most, British) voice, and despite some Ruskin scholarship in Italy and Japan (where his writings were influential in Meiji art and nature writing),[1] Ruskin studies are dominated by students of English Victorian life and letters.[2] Focusing on Ruskin's geographical imagination allows us better to understand how his work actually draws together uniquely English perspectives and concerns with a profound sense of the importance of England's relationships with a broader world, in continental Europe and beyond. As Giorgio Mangani has recently pointed out, the initiating purpose of geographical knowledge and representation is to 'make visible' the world beyond immediate experience.[3] In doing so geographers have always used images, in written description, pictures and maps. In fulfilling his self-appointed mission to bring the cultural history of continental Europe to bear upon the condition of nineteenth-century England (however distinctly he viewed the island's history and social development) Ruskin's principal techniques of description, imagery and indeed mapping were distinctly geographical. In this chapter I address Ruskin's ideas about Europe and the ways he repeatedly constructed geographies of the continent in order to shape arguments directed to Victorian Britain.

'Europe' and 'European' are of course not natural geographic or ethnographic categories; they are contingent and historical constructions. This is especially true for Europe as a geographical space: of all the 'continents', Europe is the most mythical.[4] It is equally true of Europe as a social and historical phenomenon: its derivation from the prior idea of 'Christendom' has been much studied.[5] When Ruskin speaks of Europe he is generally referring to a highly restricted geographical area delimited by his personal travel experience. But, as we shall see, his imaginative and rhetorical geography of the 'Continent' can vary considerably: sometimes treating it as a single cultural entity against which England or Britain is contrasted, sometimes focusing on its diverse and conflicted 'nations', sometimes making Europe stand for the whole 'globe of earth', across which world historical forces contend and collide. Throughout, Ruskin remains very distinctly a Victorian Briton, constrained in many respects by the prejudices of his native island and devoted to its moral, social and environmental welfare. While fascinated by what he takes to be Europe's otherness, and occasionally engaged or enraged by its politics, his soul assuaged by its scenery and the romance of its picturesque landscapes, Ruskin rarely engaged with the contemporary life and struggles of nineteenth-century Europeans. His was a geographical vision shaped in large measure from beyond its object of study, and in his final years Ruskin retreated physically and emotionally into the most English of landscapes, along the shores of Coniston Water.

My discussion is structured by the familiar trajectory of Ruskin's own life and thought: from his early engagement with a picturesque art in which landscape and scenery are important points of localized cultural expression; through his pursuit of Turner's romantic vision into the Alps and to Venice, and his discovery there of an architectural form that Ruskin believed spoke directly to the social malaise of industrial England; and finally to his deepening pessimism about a contemporary world darkened by the storm clouds of carboniferous manufacturing in Britain and by growing conflict between the European nations.[6] As any student of Ruskin knows, there are no coherent chronologies or consistencies to be found within Ruskin's thinking and writing. In deference to this fact I shall draw from across the body of his work, in order to limn as cogent a picture as I can of John Ruskin's European geography.

Grand tourist in picturesque Europe

No one has ever matched John Ruskin's ability to put into words the qualities that early nineteenth-century picturesque painters and engravers produced in sketch and watercolour, copper and steel engraving. Here, for example, he describes from a diary entry of 1854 what for a Victorian arriving from England would have likely been among the first scenic visions of the continent:

> I cannot find words to express the intense pleasure I have always in first finding myself, after some prolonged stay in England, at the foot of the old tower of Calais church. The large neglect, the noble unsightliness of it; the record of its years written so visibly, yet without sign of weakness or decay; its stern vastness and gloom, eaten away by the Channel winds, and overgrown with the bitter sea grasses; its slates and tiles all shaken and rent, and yet not falling; its desert of brickwork full of bolts, and holes, and ugly fissures, and yet strong, like a bare brown rock; its carelessness of what any one thinks or feels about it, putting forth no claim, having no beauty or desirableness, pride, nor grace; yet neither asking for pity; not, as ruins are, useless and piteous, feebly or fondly garrulous of better days; but useful still, going through its own daily work, as some old fisherman beaten grey by storm, yet drawing his daily nets: so it stands, with no complaint about its past youth, in blanched and meagre massiveness and serviceableness, gathering human souls together underneath it; the sound of its bells for prayer still rolling through its rents; and the grey peak of it seen far across the sea, principal of the three that rise above the waste of surfy sand and hillocked shore, the lighthouse for life, and the belfry for labour, and this for patience and praise.[7]

We might be standing before a rendering by Samuel Prout or Copley Fielding, those early nineteenth-century painters of picturesque landscape who had acted as drawing masters at different times to Ruskin, or indeed before one of Ruskin's own early images, modelled on their styles. But, while picturesque art in Regency England largely served the decorative demands of a growing bourgeois market, destined for the walls of newly built villas in the suburbs of London, Birmingham or Manchester, Ruskin was acutely aware of the moral debates that had underpinned picturesque theory from its enunciation by William Gilpin, Uvedale Price, Richard Payne Knight and Humphry Repton.[8] Thus, his consideration of Calais church quickly turns to the lessons to be taken home from the pleasures of witness on the French coast:

> I cannot tell the half of the strange pleasures and thoughts that come about me at the sight of that old tower; for, in some sort, it is the

epitome of all that makes the Continent of Europe interesting, as opposed to new countries; and, above all, it completely expresses that agedness in the midst of active life which binds the old and the new into harmony. We, in England, have our new street, our new inn, our green shaven lawn, and our piece of ruin emergent from it, a mere *specimen* of the Middle Ages put on a bit of velvet carpet to be shown, which, but for its size, might as well be on a museum shelf at once, under cover. But, on the Continent, the links are unbroken between the past and present, and, in such use as they can serve for, the grey-headed wrecks are suffered to stay with men; while, in unbroken line, the generations of spared buildings are seen succeeding each in its place. And thus in its largeness, in its permitted evidence of slow decline, in its poverty, in its absence of all pretence, of all show and care for outside aspect, that Calais tower has an infinite of symbolism in it, all the more striking because usually seen in contrast with English scenes expressive of feelings the exact reverse of these.[9]

That sense of rupture and loss that accompanies modernity, and to which some observers have attributed the invention of nostalgia in the later eighteenth century, is fundamental to the picturesque sensibility.[10] For Ruskin, England represents a 'new' country, not in the sense that he regarded America as new, and therefore incapable of producing landscape because it lacked the historical depth of human (ie European) settlement, but because England's modernity has produced a disjuncture between its history and its present, and that disjuncture is visibly apparent in its landscape. Europe, by contrast, he treats as a repository of historical continuities, present not because of some moral superiority of French, Italian or German peoples, but because they are less advanced along the path of modernization: 'newness' in Ruskin's terminology.

To understand his argument and its implications we need to consider the physical and intellectual environment in which Ruskin was raised. Born in 1818, he spent an extended childhood in the newly constructed suburbs of south London. Herne Hill and Denmark Hill did indeed represent radically new landscapes. Set in their own grounds, individual villas parodied the genteel country house of the eighteenth-century aristocracy, but they surveyed no rural estate, and ranged along urban streets and avenues. The English suburb had no historical precedent and no geographical parallel at this time. As such, and in the social context of England before the Reform Bill, the suburb was a space of intense insecurity. Strongly gendered, tightly stratified and uncertain of its public face, the middle-class suburb found stability in decorum. Its residents thus offered a market for improving texts and advice on manners, such as that contained in Mrs Beeton's *Book of Household Management*. John Claudius Loudon, whose *Encyclopaedia of Gardening* and other writings on the proper design and management of the villa garden, would publish John Ruskin's first essays.[11] Loudon's publications catered explicitly to the new middle class, dealing precisely with questions of the appropriate design of their suburban landscape.

But south London was only one facet of Ruskin's early formation. Equally significant was his mother's intense evangelical faith and consequent instruction of her son through daily recitation of the Testaments – to the point that biblical references inform his every text, and the Bible's cadences structure his very mode of prose composition. His education and early life were intensely domestic, and Ruskin would carry that domesticity with him into Europe. Ruskin was not one of those Englishmen for whom Europe's appealing otherness was to be found in an exotic street life, café culture and public spaces (unlike Ruskin's wife Effie, for whom Venice's appeal was very much 'society'). On the other hand, Ruskin's father, whose fortune he would inherit, was in intimate and daily touch with worlds that stretched beyond the suburb and onto the continent. The Ruskin family wealth was made in the sherry trade, in partnership with the elder Ruskin's Spanish business partner Domecq, and the firm's best customers were drawn principally from the gentry and rural landowning class of England. Ruskin recalls in his autobiography, *Praeterita*, the pleasure of early carriage tours with his father, visiting the more substantial villas and country houses whose designs and pretensions Herne Hill homes aped. In these wealthy houses, the memory of the eighteenth-century European Grand Tour was often visible in paintings on the walls and prints in the library.

The old tradition of aristocratic grand tourism had died with Napoleonic Europe, or rather, it became more demotic and extended horizons further afield into Greece and the Levant. When Englishmen could again freely travel on the continent, about the time of Ruskin's birth, both Europe and they had changed. Middle-class English people could now afford to follow the routes previously traced by aristocratic tourists, and soon would do so by train rather than private carriage. And they journeyed secure – even contemptuous – in the knowledge of their own nation's military, industrial and financial superiority. The Ruskins took early advantage of these opportunities, so that the young John experienced both the last years of guided travel by private carriage and guide, and later the coming of mass tourism, directed by the instructions of Baedeker and John Murray, whose guidebook to Normandy Ruskin himself would write.[12] The Ruskins' first European tour, undertaken when the younger John was 17, was a pursuit of Turnerian landscape. It thus traced the aristocratic grand tourist route that Ruskin's artist hero himself had followed. Thenceforth, John Ruskin's lifelong itineraries through Europe varied only in local details. Departing from London and crossing the Channel to Calais or Boulogne, he would pass through northern France (never paying serious attention to Paris), to the

Jura, the Rhine valley and the Alps, via either Fribourg and the Jungfrau, or Chamonix and Mont Blanc, thence into Italy by way of Padua or Verona. Rarely did he travel as far south as Rome, and only once to Naples, Sorrento and Amalfi. More commonly his destination was Florence or, famously, Venice. Briefly, he contemplated taking up residence in Switzerland, but his only sustained domicile on the continent was during study visits to Venice and the Gothic monuments of north Italy, which never exceeded 12 months.[13]

'The architecture of the nations of Europe'

This context helps clarify Ruskin's response to Calais church. He viewed Europe through the eyes of a leisured traveller arrived from England, and principally concerned with matters of art, architecture and landscape. Standing before the building, Ruskin's immediate point of comparison is the English suburb that was home, with its 'new street, . . . new inn, [and] green shaven lawn'. And it is this new landscape that generates the question he set himself, as a neophyte writer, fresh from Oxford, in a series of essays for Loudon's *Architectural Magazine*. These 1838–9 pieces, later collected into *The Poetry of Architecture*, examine the connections between the construction and style of domestic buildings and the natural landscape in which they are set. Ruskin opens with a criticism of the chaos of domestic architectural styles to be found in the immediate environs of the modern English city: 'We have Corinthian columns placed alongside pilasters of no order at all, surmounted by monstrosified pepper-boxes, Gothic in form and Grecian in detail, in a building nominally and peculiarly "National"; we have Swiss cottages, falsely and calumniously so entitled, dropped in the brickfields around the metropolis.'[14] His criticism is informed by a strong picturesque sensibility. Ruskin classifies landscapes by colour: 'cultivated or blue country', 'wooded, or green country', for example, each determined by topography, sky and atmosphere, soil, vegetation and land use. Architectural focus is on constructional materials, elevation and decorative elements – chimneys, treatment of windows, doors and architraves, carving of eaves – rather than architectonics, internal plan or functional space.

Ruskin's subtitle is 'the architecture of the nations of Europe considered in association with natural scenery and national character'. It suggests a geographical rather than formal or theoretical derivation of design principles, and takes us beyond purely English picturesque into a broader debate that was widely diffused in early nineteenth-century European Romantic criticism. *The Poetry of Architecture* proposes

> to trace the distinctive characters of the architecture of nations, not only its adaptation to the situation and climate in which it has arisen,

but its strong similarity to, and connection with, the prevailing turn of mind by which the nation who first employed it is distinguished.[15]

Belief in a relationship between a 'nation,' its physical geography and its art, was widespread in post-Napoleonic Europe, and widely used as an ideology of nationhood. Germany offers the clearest example, as the disciplines of geography and art history developed there in close parallel, although similar connections are found in France and Italy.[16] We can find parallel studies to Ruskin's in the writings of both Domenico Fiorillo, holder of the first Chair of Art History at the University of Göttingen from 1813, and Carl Ritter, Chair of Geography in the same institution in 1825, whose work included studies of Gothic architecture. Belief in the close connections of art, place and people was the basis for Johann Gottfried Herder's influential *Volksgeist* theories that echoed through more than a century of German *Geisteswissenschaft*. In Britain, the idea of a 'national school' of art had emerged at the turn of the century as part of the nationalist response to French expansionism, and Ruskin would make his reputation by defending the idea of a national art and especially Turner's primary place within its pantheon. Across Western Europe, art history in the early decades of the nineteenth century was caught in a contradiction of searching for national distinction rooted in geography while proclaiming participation in such (newly minted) pan-European movements as 'Gothic', 'Renaissance' and 'Romantic'. Ruskin's earliest work thus reveals his participation in a conversation that was genuinely 'European'.

Venice and European Gothic

Ruskin's vision of Europe was shaped initially and principally by J. M. W. Turner's art, which the young writer began to collect about the time he composed *The Poetry of Architecture*. It was Turner's vignettes for Samuel Rogers's romantic verse work *Italy* that introduced Ruskin to the ageing artist, and propelled the young critic to the Alps, north Italy and Venice. Turner's Venetian works are among his most revolutionary in dissolving line and form into colour and light, the very images that Ruskin sought to defend in the first volume of *Modern Painters*, published in 1843. That enormously influential text was dedicated to revealing the ways in which the 'English school' drew upon but superseded all previous art in the field of landscape, because it was supposedly rooted in the English 'mind' and English geography. At the same time, Ruskin developed some of his criticisms of the material landscapes that were appearing in mid-nineteenth-century Britain. *Modern Painters*, a *tour de force* of critical writing for a 25-year-old, made huge, and often unsubstantiated, claims about art and architecture over a span of four centuries, and across a geographical field that ranged from Britain and

Holland to Italy and Spain. Its reception, together with that of the second volume in 1846, made Ruskin a national figure, exhausted him emotionally, and forced him to recognize that if future volumes were to carry true critical authority he would have to devote himself to a much more detailed and direct study of the high Renaissance art and architecture of which Modern Painters I was so dismissive. It was this that took him to Venice in 1849, a visit delayed by a year because of the political upheavals of 1848. This first extended stay in a European city delayed the proposed further volumes of Modern Painters but opened an entirely new project: an art and architectural history of Venice.

Venice is the first of two cities that Ruskin would claim as key European centres: meeting points for great flows of cultural influence that shaped the continent's (and thus for Ruskin, the world's) culture. The other city, as we shall see, was Amiens. In both places he draws upon the same conceit: that the city's principal public monument acts as a civic 'bible,' expressing the moral principles upon which civil society rests. In Venice it is the Doge's Palace, in Amiens, the cathedral. The Stones of Venice and The Bible of Amiens construct moral geographies shaped by Ruskin's vision of Europe and Christian civilization. These are both complex and even contradictory texts and I draw from them only narrow threads that speak directly to Ruskin's concept of Europe.

By far the best-known and most influential writing in The Stones of Venice appears in the second part of Volume II. 'The Nature of gothic' drew on the first volume's analysis of Venetian architecture to address the topical debate over what constituted Britain's 'national style' of architecture. The dispute over 'Greek' or 'Gothic' had been brought into the public eye by A. W. Pugin's Contrasts of 1836, and was made especially current through the question of rebuilding the Palace of Westminster, destroyed by fire in 1834 (and memorialized in a famous painting by Turner). By the time Ruskin wrote, the Gothic style had effectively won the argument. His contribution was to connect the nostalgia that underlay picturesque taste for Gothic to a thoroughgoing critique of contemporary industrialism. To achieve this, he brought the principles of beauty, truth and imagination that he had developed in Modern Painters to bear on an analysis of Gothic architecture as a decorative form, and to relate that form to the imagined social order of medieval guild craftsmanship, which Ruskin held up as a mirror to what he regarded as the ugliness, dishonesty and mechanical alienation of contemporary Britain.

'The nature of Gothic' opens with one of Ruskin's most sustained prose poems, an imagined flight from Africa to the north polar sea that encompasses the whole European continent. Despite its length, it is worth quoting the passage at length, for it captures perfectly both Ruskin's poetic geography and his construction of a moral geography of Europe:

The charts of the world which have been drawn up by modern science have thrown into a narrow space the expression of a vast amount of knowledge, but I have never yet seen any one pictorial enough to enable the spectator to imagine the kind of contrast in physical character which exists between Northern and Southern countries. We know the differences in detail, but we have not that broad glance and grasp which would enable us to feel them in their fullness. We know that gentians grow on the Alps, and olives on the Apennines; but we do not enough conceive for ourselves that variegated mosaic of the world's surface which a bird sees in its migration, that difference between the district of the gentian and of the olive which the stork and the swallow see far off, as they lean upon the sirocco wind. Let us, for a moment, try to raise ourselves even above the level of their flight, and imagine the Mediterranean lying beneath us like an irregular lake, and all its ancient promontories sleeping in the sun: here and there an angry spot of thunder, a grey stain of storm, moving upon the burning field; and here and there a fixed wreath of white volcano smoke, surrounded by its circle of ashes; but for the most part a great peacefulness of light, Syria and Greece, Italy and Spain, laid like pieces of a golden pavement into the sea-blue, chased, as we stoop nearer to them, with bossy beaten work of mountain chains, and glowing softly with terraced gardens, and flowers heavy with frankincense, mixed among masses of laurel, and orange, and plumy palm, that abate with their grey-green shadows the burning of the marble rocks, and of the ledges of porphyry sloping under lucent sand. Then let us pass further towards the north, until we see the orient colours change gradually into a vast belt of rainy green, where the pastures of Switzerland, and poplar valleys of France, and dark forests of the Danube and Carpathians stretch from the mouths of the Loire to those of the Volga, seen through clefts in grey swirls of rain-cloud and flaky veils of the mist of the brooks, spreading low along the pasture lands: and then, farther north still, to see the earth heave into mighty masses of leaden rock and heathy moor, bordering with a broad waste of gloomy purple that belt of field and wood, and splintering into irregular and grisly islands amidst the northern seas, beaten by storm and chilled by ice-drift, and tormented by furious pulses of contending tide, until the roots of the last forests fail from among the hill ravines, and the hunger of the north wind bites their peaks into barrenness; and, at last, the wall of ice, durable like iron, sets, deathlike, its white teeth against us out of the polar twilight.[17]

Ruskin proceeds to map in similarly graphic prose the parallel bands of vegetation and animal life which correspond to this 'gradation of the zoned iris of the earth', and finally asks us to consider how the works of human art in transforming nature, or building sacred and secular structures, also submit to the same cartographic logic: 'the expression by man of his own rest in the statutes of the lands that gave him birth'.

But this geography is not static. Studying the history of Venice, Ruskin could not be unaware of the city's role as a mercantile trading centre of early modern Europe. Indeed, the declared purpose of the entire study is to present the parallels between Venice and Britain as maritime civilizations, open to influences from overseas while needing to remain true to their geographical foundations:

> Since the dominion of men was first asserted over the ocean, three thrones, of mark beyond all others, have been set upon its sands: the thrones of Tyre, Venice, and England. Of the First of these great powers, only the memory remains, of the Second, the ruin; the Third which inherits their greatness, if it forget their example, may be led through prouder eminence to less pitied destruction.[18]

Human art reflects not only its roots in a particular physical environment, but 'the degree of intellectual and moral energy of the nations which originated them'.[19] In Ruskin's view Christianity offers the fullest development of the human spirit and its civilization must give rise to a new architecture, advancing upon all previous forms. Venice represented in his view a key moment in the evolution of medieval Christian civilization, partly because of its geographical position at the juncture of Christianity's Eastern and Western expressions (themselves reflections of global cultural divisions).

> One might look over Europe and see how each town takes its natural position – and becomes prosperous if its happens to understand that position and take due advantage of it – and one might say – generally, Genoa grew up in the place for Genoa – and Rotterdam in that for Rotterdam, and Venice in that for Venice. It seems that just in the centre of Europe, at the point where the influences of East and West – of the old world and the new – were to meet, preparation was made for the existence of a city which was to unite the energy of the one with the splendour of the other.[20]

The Hegelian and Orientalist aspects of this argument hardly need rehearsing; they were part of the self-definition of Europe for nineteenth-century intellectuals across the continent. Thus Venice's European centrality implies for Ruskin a global centrality. The Doge's Palace 'contains the three elements [of the three antecedent architectures] in exactly equal proportions – the Roman, Lombard and Arab [read Byzantine]: It is the central building of the world', and the bible of the city.[21]

The century of Venetian commercial hegemony in the eastern Mediterranean between 1300 and 1500 is taken by Ruskin to correspond with the period of its greatest moral and constitutional perfection, and thus a uniquely expressive elaboration of its distinct Gothic architecture. In Ruskin's Venetian geography the Doge's Palace 'rose in the midst of other work, fanciful and beautiful as itself . . . every dwelling house in the middle ages was rich with the same ornaments and quaint with the same grotesques which fluted the porches or animated the gargoyles of the cathedral'.[22]

The implication of Venice's moral geography and its subsequent constitutional and moral decadence for contemporary Britain is not to do primarily with design; indeed Ruskin would later bemoan the influence of his works 'on nearly every cheap villa builder between this and Bromley; and there is scarcely a public house near the Crystal Palace but sells its gin and bitters under pseudo-Venetian capitals copied from the church of the Madonna of Health or of Miracles'.[23] Rather, the moral Ruskin took for Britain from Venice's experience as the centre of Europe concerned enduring lessons of social organization and moral order; their material expression was consequent to the organization of life and labour, a belief he would seek to put into practice, with decidedly mixed results.

The Bible of Amiens

It is needful evidently that we should have truthful maps of the world to begin with, and truthful maps of our own hearts to end with . . .[24]

'The nature of Gothic' signalled John Ruskin's embrace of passionate social and political engagement. While he wrote and lectured on art for the remainder of his life, the activist intent of his work was always explicit and often determinative. And with age and mental ill-health Ruskin's social vision grew increasingly declensionist. His sustained fascination with Gibbon's history of imperial Rome's closing years reflects the map of Christendom that continued to shape his geographical imagination. In the numerous cartographies of artistic expression, cultural change and moral geography that served his arguments, the world's axis for Ruskin would continue to extend from Sicily to the English Lakes and Scottish Highlands.

Ruskin's European geography receives its last and most sophisticated elaboration during the last years of his writing, in The Bible of Amiens, published in 1885. As I have shown in the previous chapter, he opens the text as an English tourist alighting from the Calais–Paris train to visit the cathedral square of Amiens and asking why such a wondrous monument should have been constructed in what Ruskin claims is 'the central square of a city which was once the Venice of France' (see Fig. 7.2). The question can only be answered through an understanding of the ways that Christian belief has woven itself through the historical competition of European nations, each expressing simultaneously its faith and the geographical conditions of its existence through its sacred architecture and sculpture. It is in the Gothic churches of Caen and Palermo, Orvieto and Florence, Gloucester and Paris, that Europe's cultural history and geography are to be read. And 'the simplest, completest, and most authoritative in its lessons to the active mind of North Europe, is on the foundation stones of Amiens'.[25]

Ruskin's analysis proposes Amiens as an *axis mundi*, a spiritual centre predating Venice where, in the decaying years of Roman imperial rule, Christianity was lodged in the humble faith of the northern, barbarian races. Amiens, he claims, lies at the meeting point of the two core nations of northern Europe, Gauls and Germans. His reasoning draws on two classical theories: the Greek concept of climatic zones and the Roman idea of *nationes*, groups sharing broad ethnic, linguistic and cultural similarities who occupied equally broadly defined territories within the empire. In Ruskin's scheme nations divide into two groups. Those 'fixed in habitation of the temperate zone of Europe . . . "Britain", "Gaul", "Germany" and "Dacia," ' Ruskin calls 'resident' races because they cultivate fields and tend orchards. To the north, stretching from the mouths of the Rhine to the Vistula, are 'a scattered chain of gloomier tribes, piratical mainly', who are nomadic, inhabiting a wet, cold and gloomy landscape of glaciated rock, heath and bog.[26] This is a reprise of the geography earlier outlined in his passage from 'The nature of Gothic'.

Ruskin's argument is rooted in a Gibbon-esque construction of European history as the ordained outcome of the encounter between Christianity and the classical world. As in the earlier text from 'The nature of Gothic', Ruskin develops his argument with reference to a map, in this case one taken from Edwin Creasy's recently published *History of England*. He devotes the early part of his text to the issue of mapping cultural history, noting that the world's continents and oceans are not only fully charted, but are increasingly interconnected by telegraph cables and steamship lines. However, Ruskin claims, the great historic question about the globe does not depend upon modern maps. It 'is not how [the earth] is divided, here and there, by ins and outs of land or sea; but how it is divided into zones all round, by irresistible laws of light and air'.[27] In other words, it is to the classical Greek *klimata*, grouped into latitudinal frigid, temperate and torrid zones, that we should look in order to interpret the globe's moral geography, a long-standing geographical concept as we have repeatedly noted. The space of history in Ruskin's view takes place within the longitudinally delimited part of the temperate zone consisting of peninsular Europe and the Mediterranean. This constitutes 'the small, educationable, civilizable, and more or less mentally rational fragment of the globe', to be differentiated from Asia's 'great Siberian wilderness . . . inconceivable, chaotic space'. Within this fragment are 'twelve countries, distinct evermore by natural laws, and forming three zones from north to south, all healthily habitable'. Towards the frigid zone he locates four *Gothic* nations: Britain, Gaul, Germany and Dacia; towards the temperate zone, the *Classic* nations of Spain, Italy, Greece and Lydia (Asia Minor); while on the southern shores of the Mediterranean, near the torrid zone, are found the *Arab* nations of Morocco, Libya, Egypt and Arabia. Distinct from all these is the chosen space of revelation: 'the small hilly district of the Holy Land', within which Jerusalem stands as the spiritual centre of all the nations.[28]

According to Ruskin, each of these territories, regardless of the contingencies of history and political control, is subject to 'eternal laws enforced over it by the clouds and stars', to which the social order of its inhabitants should correspond.[29] As he had argued more generally in 'The nature of Gothic', that order will be expressed most truly in human transformations of the immediate natural world through art and architecture. Thus, aesthetic judgements of the arts of a nation become judgements on its moral condition, and vice versa. As the disciplined expression of faith, art reaches its highest expression in the Christian space that emerges among the *nationes* formerly gathered into the Roman empire. And the most energetic of these at a given historical moment achieves in its public architecture truthful expression of virtue. Historically, the great exemplars are thirteenth-century Amiens and fourteenth-century Venice. In the latter, as we have seen, Byzantium and Rome meet and coalesce; in the former it is Gaul and Germania. Eventually, the great Gothic heritage of the Île-de-France would become the cultural heart of the modern French kingdom (Fig. 8.1).

In this construction, history becomes a moral narrative of the rise and fall of different European peoples in a continuing struggle to preserve civilization, which is figured as a torch carried from Greece to Rome, confirmed in Christianity, handed to medieval burghs such as Venice and Amiens, and now passed to Britain. Civilization's 'other' is barbarism, initially located in Europe's north, but most permanently to be found looming from the eastern steppes. In *The Bible of Amiens* Ruskin writes of the Vistula and the Dneister as 'a thousand miles of moat, separating Europe from the Desert, and reaching from Dantzic to Odessa'.[30] The moat extends south against an explicitly defined Orient that lies beyond the eastern *limes* of the Roman empire:

As the northern kingdoms are moated from the Scythian desert by the Vistula, so the southern are moated from the dynasties properly called 'Oriental' by the Euphrates . . .

This valley in ancient days formed the kingdom of Assyria as the valley of the Nile formed that of Egypt. In the work now before us, we have nothing to do with its people, who were to the Jews merely a

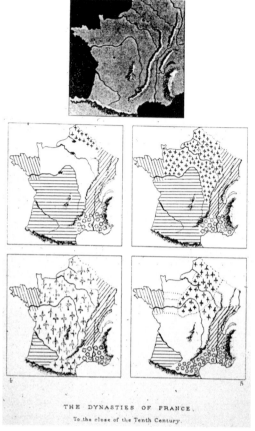

THE DYNASTIES OF FRANCE.
To the close of the Tenth Century.

Figure 8.1 John Ruskin, Map of the kingdoms of France. 'Sketch for yourself, first, a map of France, as large as you like . . . Fig.1, marking only the courses of the five rivers, Somme, Seine, Loire, Saone, Rhone; then, rudely, you find it was divided at the time thus, Fig.2: Fleur-de-lysée part, Frank; \\\, Breton; ///, Burgundian; =, Visigoth.', from 'The Bible of Amiens', *The Works of Ruskin*, 33: 34 (author's copy)

hostile power of captivity, inexorable as the clay of their walls, or the stone of their statues; and, after the birth of Christ, the marshy valley is no more than a field of battle between West and East.[31]

This is a geographical construction whose origins date back at least as far as Strabo.[32] But beyond Ruskin's classical learning and medievalist fantasy *The Bible of Amiens* betrays a core of recognizably late nineteenth-century British geopolitics. The text was written in 1880, in the interval between Ruskin's two periods of art teaching at Oxford. Four years later, Europe's political hegemony over the globe would be confirmed by the Berlin

Plate 1 Peter Bruegel the Elder, *The Fall of Icarus*, c.1558 (Musée des Beaux Arts, Brussels)

Plate 2 Detail from Adam Gray, *Microbial*

Plate 3 Michael Light, remastered NASA photograph of astronaut Alan Bean walking on the Moon during the Apollo 12 mission, 1969 (Michael Light, *Full Moon*, 1999)

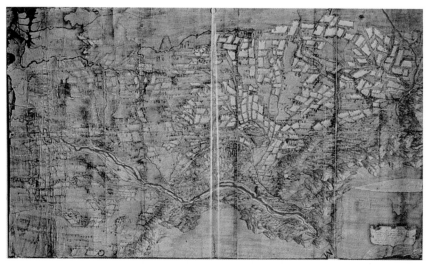

Plate 4 Cristoforo Sorte, Reclamation map of the Piave valley, 1580 (Archivio di Stato di Venezia)

Plate 5 John Ruskin, *Fribourg*, watercolour, 1859 (Teaching Collection, Ashmolean Museum, Oxford)

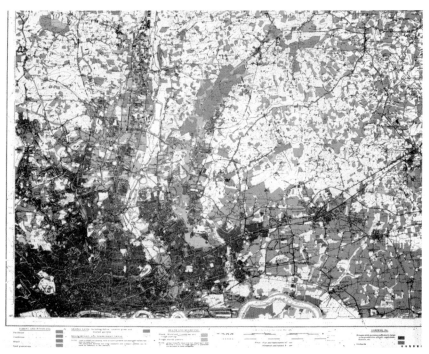

Plate 6 London region, National Land Use Survey map: 1:63,360, *c.*1938 (Ordnance Survey, Southampton)

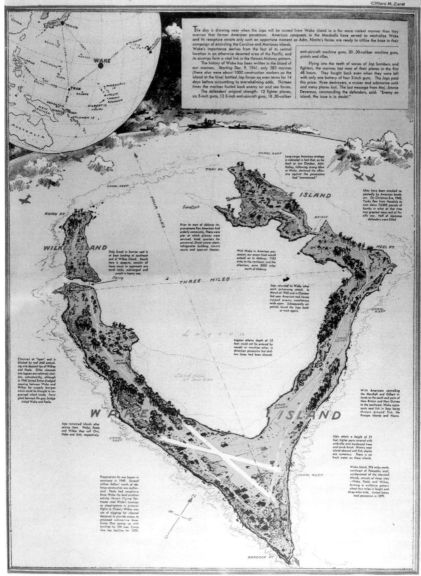

Plate 7 Charles H. Owens, Wake Island, *Los Angeles Times* war map, 1944 (UCLA Library)

Conference's division of colonies between the colonial 'great' powers, and the Conference of Washington's confirmation of Greenwich as the prime meridian of the globe. Europe's hegemony was being justified by the arguments of scientific racism and environmental determinism that continued to map onto the simplicities of climatic zones (discredited by Humboldt in the early years of the century, as Ruskin well knew), a hierarchy of peoples and civilizations ranked in descending order by their distance from Europe. In fact Ruskin made use of a crude woodcut representation of racial types in his Oxford art classes, in order to illustrate the lack of artistic merit in popular illustration. Imperial Britain, whose Anglo-Saxon population figured itself at the very peak of this environmental and cultural hierarchy, looked east, to a dubiously European Russia, for the greatest threat to its global supremacy.

In 1884, a year before *The Bible of Amiens* appeared, Ruskin wrote his last influential work, the deeply pessimistic *The Storm Cloud of the Nineteenth Century*, in which the moral and social decay of contemporary England is read in the climatic evidence of a series of cool, cloudy summers and grey winters that had prevented Ruskin from surveying and sketching his beloved Lakeland landscapes. Rejecting industrial air pollution as the sole cause of the storm cloud, or plague wind as he also calls it, Ruskin comments that it 'looks more to me as if it were made of dead men's souls'.[33] He continues with an aside about the deaths caused by the Franco-Prussian War that had been fought in the very months that he was writing his text on Amiens. That war, Ruskin claimed, 'was especially horrible to me, in its digging, as the Germans should have known, a moat flooded with the waters of death for a century to come'.[34] The Franco-Prussian War was the first major international conflict to take place on European soil during his own lifetime, and it opened, as he predicted, a 70-year struggle between the two nations whose destinies he had woven together in *The Bible of Amiens*. In the decades following his death, their struggle would be played out across fields of slaughter, 'on the chalk and finely-knit marble, between . . . Amiens and Chartres one way, and between Caen and Rheims on the other . . . real *France*', the France that the founder of the French school of regional geography, Vidal de la Blache, and his students were celebrating in monographs written in the very years that Ruskin was composing his last works. For Ruskin's own countrymen, it was to be at Amiens that so many thousands would perish, in 'moats flooded with the waters of death', and in a landscape reduced to an elemental chaos of earth and putrid water, smoke-laden air and shellfire.

Conclusion

Today, Ruskin's significance as a Victorian European does not lie in the prescience of his writings about warfare on the chalk plateaus of northern France, so much as in his grasp of some of the thorniest questions of culture, space and environment within today's Europe. We may no longer feel comfortable with the evangelical language in which his analysis is cast, but we cannot avoid the continued, even growing significance for the twenty-first century of some of the questions his writings on Europe addressed in the nineteenth.

Today's European Union was born out of the experience of twentieth-century warfare, and especially the post-war desire for a settlement of the Franco-German conflict, whose outset Ruskin had so ruefully witnessed. Now in its sixth decade, the principal philosophical questions facing the Union concern the framing of a common constitution, and defining Europe's boundaries. The constitutional question depends politically on how far different 'nations', whose distinctions can be traced or constructed historically back to the Roman empire, may be moulded into a single polity. Philosophically, that question turns on the values these 'European' nations hold in common. For Ruskin the answer to the question of Europe's boundaries would have been clear: they are defined by Christianity – the commonalities of art and architecture, above all the Gothic, derived from a shared faith. Thoughtful commentators today who stress the common heritage of art and architecture (Gothic, Renaissance, baroque) as a foundation for European consciousness, are understandably reluctant to raise the spectre of faith as a common foundation in a multicultural and largely secular Europe. Thus, to the chagrin of some observers, including expressly the (German) Pope Benedict XVI, the European constitution upon which member nations voted in 2005, with such dismal results for its writers, contained no reference to Christianity. Were it to do so, any possibility of welcoming within the union a nation such as Turkey – Lydia within Ruskin's geography – would necessarily have to be abandoned. Ruskin's analysis might offer some clues as to why such an outcome seems to find more favour in contemporary France and Germany than in Britain.

The contradictions inherent in any attempt to link people, territory and cultural expression are unavoidable. Thus, while cultural study today is eager to embrace geography as a source of theoretical insight, the attention is directed towards questions of space and identity rather than environments and peoples. The relations between place, territory and artistic expression that Ruskin and his continental contemporaries sought to identify and evaluate across Europe lie largely discredited, principally because the worst excesses of twentieth-century European nationalism were predicated upon their proclaimed essentialism. But as the Princeton art historian

Thomas da Costa Kaufmann concludes in his survey of the disciplines of art history and geography, their intimate association over the past two centuries reflects the inescapable fact that works of art are human products, and as such 'provide evidence of human culture, specifically the culture of whatever people produced them'.[35] Like the constitutional debates within the European Union, the possible revival of *Kunstgeographie* in continental European scholarship today is a reminder that, while the answers which a Victorian European such as Ruskin so confidently offered may have been misguided, the questions he asked still remain to be answered.

V Cartographic visions

9 Moving maps

The historiography of maps and mapping – of 'cartography', to use the nineteenth-century neologism – has been revolutionized in the past quarter-century. The shift is as profound as the changes that have taken place in map-making and map use themselves, resulting from satellite remote sensing, digitization of spatially referenced data and computer manipulation of information and images. Three main changes in cartographic historiography are worth highlighting. First, there has been detailed exposure of the normalizing and often ideological authority of maps, and criticism of the active roles cartography has played in the nexus of power-knowledge that frames and shapes the geographies of the modern world. Geographic and topographic mapping and maps have been critical tools for the modern state and its agencies in shaping social and moral spaces, and they played a central role in the Western physical and intellectual colonization of territories, peoples and the natural world. For over two centuries thematic and statistical maps have extended these roles in supporting the bureaucratic concerns of the modern state. Historians of cartography have examined these processes across many specific instances.[1] Second, map-making's scientific claims to offer progressively accurate and objective, scaled representations of spatial relations, have been challenged with recognition of the inescapable imaginative and artistic character of cartographic process and products that accompany framing, selection, composition and graphic representation of mapped information. Colour and symbolization, for example, are chosen and applied to maps according to widely accepted design principles, but their relationships to the appearance of landscapes represented on topographic maps, or the bands within the infrared spectrum on a remote sensed image, are necessarily arbitrary. This recognition has opened up an exciting new field of connections between scientific map-making and creative art practices.[2] Third, and closely related to the first two developments, is the recognition of mapping as a complex cultural process

155

in which the map itself represents merely one stage. To understand the contents, meaning and significance of any map requires that it be reinserted into the social, historical and technical contexts and processes from which it emerges and upon which it acts. This involves examining the map not only as a discrete object but as the outcome of specific technical and social processes and the generator of further social processes as it enters and circulates in the social world.[3]

These assumptions of the critical approach to the nature and history of maps are today widely accepted among historians and writers on cartography, but many would argue that they are still insufficiently embraced by practising map-makers, especially today when many of those producing and manipulating maps are not trained cartographers. Most professional cartographers are acutely aware of the limits of their art even as they strive for disinterested objectivity and scientific integrity in their map-making.[4] But maps made by formally trained cartographers constitute an ever-smaller proportion of the map images available today, especially those available online rather than drawn or printed. Geographic Information Science (GIS), working with remote sensed digital data at intensive scales and across a colour spectrum that stretches deep into the infrared regions, generates a vast range of virtual and actual cartographic images. Their makers' primary training and interests are often in information technology and its applications to geo-referenced data rather than in the conventional cartographic techniques and operations of projection, compilation and selection, framing and design.[5] The sheer technical wizardry and compelling graphic effects of such animated mapping packages as Google Earth offer an illusion of total synopsis and truthful vision, and can easily blunt critical responses and obscure significant continuities in cartographic culture.[6] In this chapter I explore some of those continuities, drawing on the critical cartographic literature to comment on aspects of maps and mapping practices that can easily become obscured in our excitement with the technical advances that have made 'mapping' such a dynamic contemporary field of practice and study.

Directions in critical cartography

Critical study of cartography can proceed from two directions: either through study of the finished map – judging its function, technique, aesthetics and semiotics; or through a study of mapping processes, conventionally grouped under the headings of survey, compilation and design. From the first perspective we might consider Abraham Ortelius's *Theatrum Orbis Terrarum* of 1570 (Fig. 9.1). Functionally, this well-known historical map provided what is considered the first modern atlas with an opening image

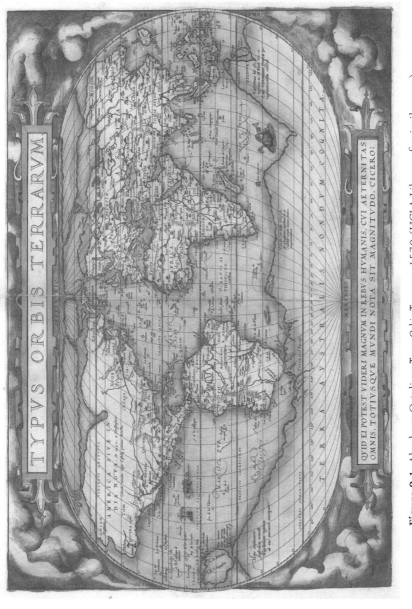

Figure 9.1 Abraham Ortelius, *Typus Orbis Terrarum*, 1570 (UCLA Library, facsimile copy)

of the terraqueous globe according to the most recent information available at the time of its making.[7] The search for empirical truth is apparent from the second edition of the map, made a mere decade later when, among other changes, the shape of South America is more accurately portrayed. Ortelius's selection of individual colours for the continents anticipates their representation on succeeding continental maps, and the ordered summary of geographical knowledge that constitutes the atlas. Technically, the world map uses Ptolemy's second projection, extending the meridians to show the whole southern hemisphere. The map is thus centred on the Equator, with a prime meridian running through the Azores, curving the longitudes towards the poles. Like any projection of the sphere, this has distorting effects on shape and direction. The oval planisphere is framed with clouds that represent the element of air, but otherwise it is relatively free of decoration, apart from the title cartouche and a lower banner containing a Latin sentence attributed to Cicero. The map offers a memorable and uncluttered image of the globe's lands and seas to which subsequent maps in the atlas can be related. Yet in ways that are not immediately apparent on the map's surface, aesthetics could be said to trump scientific knowledge in the balancing landmasses north and south of the known continents: remnants of philosophical and religious belief in a harmonious distribution of lands over the earth's surface. The semiotics of the map are as significant as its scientific, technical and aesthetic aspects. The text at the base of the map, for example (which in the second edition is reinforced by four other passages from Cicero and Seneca), reads 'For what can seem of moment in human affairs for him who keeps all eternity before his eyes and knows the scale of the universal world?' It reminds us that in the sixteenth century the world map played a role beyond that of scientific instrument and artistic image; it was a moral text reminding the viewer of the insignificance of human life compared to the vastness of creation.[8] In presenting the mapped 'theatre of the world' (the title of Ortelius's atlas) as a moral space, the map itself gains an emblematic quality.[9] This aspect of mapping can be traced in the West back to the medieval Christian *mappae mundi*, and is a common feature of non-Western cartography too.[10] Indeed it has never disappeared from cartographic culture.

The alternative approach to a critical understanding of the cartographic image is via an examination of survey and compilation, and these will be my focus here. By survey I mean the direct collection and production of the spatial data to be represented, or 'mapped'. This includes both the spatial calculation used to create a base map, for example a local traverse or a geodetic measure, and the informational content to be represented.[11] Survey, or reconnaissance, has traditionally been a field-based activity, within which instrumentation has played a formative role. By compilation I mean the gathering together of surveyed information at a single location –

the cartographer's office, laboratory or studio – and its technical transformation into the finished map object. This approach privileges the process of mapping over the product, the map, and it has attracted increasing attention in recent years, especially as its processes impact upon the knowledge claims that might be made of the resulting map.[12] Of particular concern have been the various means whereby knowledge gained in survey is transferred back to the place of compilation, and whereby the map itself enters recursively into circuits of knowledge that generate further mappings.[13] Mapping is itself a spatial process that involves negotiating various aspects of securing and maintaining the integrity of cartographic data as they circulate in space.

Survey

Survey is an embodied process involving direct, sensual contact with the spaces to be mapped. The distanced sense of vision is privileged here, as the word itself implies (Latin: *super-video*), although other senses can be critical in specific mapping situations, such as the whole body engagement involved in plotting precise locations or fixing trigonometric survey points in the field, or in negotiating darkened spaces for underground mapping, where bodily touch is more significant than sight.[14] Historically, there has been a progressive shift away from the individual human body as a reliable agent for recording spatial information, towards dependence upon instrumentation as the guarantor of accuracy and objectivity in survey data. This is apparent in the use of compass, astrolabe and cross staff, later alidade and lens-based instruments, and most recently light-sensitive remote sensing aids to the human eye. Optical instruments not only extend the scope of human vision; historically they have been used to supplant it. Thus Galileo's revolutionary mappings of celestial movement and imperfections on the surface of celestial bodies were founded on the capacity of the telescope lens not only to reveal the surface corrugations of the moon (traced directly onto paper by Galileo's hand), but to allow the sun to burn the pattern of its dark spots directly onto the paper with no apparent human intervention.[15] The eighteenth century saw the radical extension of instrumentation to all aspects of reconnaissance: the alidade and plane table, as well as accurate geodetics, made possible the production of national maps based on triangulation, a technique that had been theorized over 200 years earlier by Gemma Frisius. In the same years the mercury barometer was used to measure and plot altitude, while accurate mapping of oceanic space only became possible with John Harrison's 1780s invention of the chronometer, which allowed relatively easy and secure measurement of longitude at sea.[16]

But despite increased reliance on instrumentally measured survey, the human eye has remained a crucial element of mapping and the use of maps. Into the twentieth century, the British Admiralty required its officers, including the most junior, to learn accurate sketching as a means for gathering and recording information about coastlines and harbours, considering drawing to be superior to any written account. Sketches were of two types: the memorial sketch, 'a delineation of a harbour, or any part of a coast, from the memory only, without notes or any immediate sight', conveying 'the general area of a bay, harbour, or island . . . shewing that some such places are there', and the eye-sketch, 'done by the eye at one station, without measuring distances; and drawn according to the apparent shape and dimensions of the land'.[17] The invention of photography and the use of balloons, followed by powered flight, furthered the displacement of the human eye in geographic and topographic mapping. In 1915 the German Oskar Messter invented an airborne automatic camera that could film a 60km by 2.4km strip of the earth's surface in a sequence of overlapping frames, to be either printed as they were or used as raw visual data. Aerial photography thus replaced to some degree the epic work of Cassini or Everest in surveying great arcs of meridian from Dunkirk to Perpignan and Bangalore to Delhi respectively.[18] Photogrammetric survey was used in mapping great colonial stretches of Africa, Australia and Antarctica into the 1950s. But even aerial photography and its contemporary successor, remote sensed imaging from orbiting satellites far above the earth's surface, have not wholly replaced the sensing human body. 'Ground truthing' was crucial for removing the errors caused to the 1950s British Antarctic Survey by magnetic deviation, cloud cover and distance distortions in polar regions.[19] It remains necessary to ensure the instrumental accuracy of remote sensed maps today.

In the mapping process, the increased accuracy and consistency of survey results secured by the replacement of the sensing but subjective human body by instruments are always threatened by the problems of transferring recorded data from the place of survey to the place of compilation. The sketch-map can play a role in this, but as the name suggests, it lacks the authority of the 'true', surveyed map. The issue of securing the accuracy of mobile knowledge is beautifully expressed in Le Petit Prince by the French writer (and early denizen of aerial survey in French colonial Africa), Antoine de Saint Exupéry.[20] His eponymous hero flies from planet to planet absorbing moral lessons from their inhabitants and discovering the strange habits of the adult world. The most beautiful of all the planets the Little Prince visits is occupied by a single old man, seated at a desk and inscribing information into a great book. He is a geographer, but he claims never to have seen the beauty of his planet. Bemused, the Little Prince asks why. The geographer responds as follows:

'The geographer is not the one who counts the towns, rivers, mountains, seas, oceans and deserts. The geographer is much too important
to go wandering about. He never leaves his study. But he receives
explorers. He interrogates them and notes their records. And if the
records of one of them seem interesting the geographer makes an
enquiry into the explorer's moral character.'

'Why is that?'

'Because a lying explorer would have catastrophic consequences for
geography books [read maps]. The same is true for an explorer who is
a drunkard.'

'How come?' said the Little Prince.

'Because drunkards see double. So the geographer would note down
two mountains where only one exists.'[21]

I shall return below to this question of securing the truth of survey
knowledge as it travels over space.

Careful instrumentation, highly regulated recording procedures, and
learned sketching techniques have all been deployed to overcome the
subjectivity of embodied observation, but its removal is never complete.
Cartographic instruments themselves have to be tested and calibrated: James
Cook's Pacific navigation was partly intended to test Harrison's chronometer. And during the seven-year survey by Pierre Méhain and Jean-
Baptiste Delambre, begun in 1792 to calculate the precise length of an arc of
meridian in order to determine the length of the metre as universal measure
(determined objectively as one forty-millionth of the earth's polar circumference), 'every page of the expedition's record was signed by each member
of the expedition or by outside witnesses to certify that the recorded meas-
urements had been performed as described. No subsequent changes were
permitted . . . the signatures radically transformed the status of the
document.'[22]

Compilation

Compilation brings its own interruptions to the apparently smooth transfer
of spatial information from the territory to the map. Reflecting anxiety in
some measure over these problems, the history of cartographic design has
conventionally been told as a transition from art to science, a progression
from the pictorial style that we associate above all with baroque maps, to the
unornamented 'plain style' of graphic presentation in the eighteenth
century. Evidence for this progression was to be found in the changing
appearance of maps themselves: the removal of cartouches, elaborate lettering and extraneous information and marginalia, and the systematic use of

non-pictorial cartographic signs. Removal of such pictorial elements works alongside other technical devices: the graticule and grid as the map's controlling spatial metrics, the removal of hachures and relief shading in favour of measured contours and spot heights to indicate topographic relief, the appearance of mathematical and numerical information (compass bearings and geodetic information, dates of survey and publication, scale, etc.) in the map's margins, and the insertion of a formal key, explaining and controlling the use of colour and symbols. All these act as much to demonstrate the scientific credentials of the map's compilation as to provide for its actual use. But we know that many such compilation decisions are inevitably arbitrary or driven by quite other than scientific considerations: why should water be coloured blue rather than green or turquoise, or lowlands coloured green or roads red? Why does the United States Geodetic Survey's topographic map indicate schools, but not the religious denomination of cemeteries or the population size of municipalities indicated on the topographic maps prepared by the French Institut Géographique Nationale? Why do British Ordnance Survey maps mark and differentiate archaeological sites by Roman and Gothic lettering? Even the remote sensed image is a product of colouring choices applied by the map-maker to pixels received by the cartographic studio in numerical, digitized form, as is apparent when one moves across the virtual surfaces of Google Earth. The colouring of the 1997 map of ocean temperatures that made the El Niño phenomenon so graphically compelling was inspired by the desire for graphic impact rather than scientific objectivity.

The intimate relations between mapping and science were initially forged in the late eighteenth and early nineteenth centuries. Enlightenment passion for universal measure and objective precision found expression in statistics, and statistics had their greatest social impact through graphic expression in graphs, charts and maps. The thematic map is an invention of this historical moment. Although William Playfair had pioneered 'lineal arithmetic', Alexander von Humboldt is credited with producing the first isoline map in 1817 (Fig. 9.2). From the mid-eighteenth century, the shaded, or choropleth map, which uses territorial boundaries as containers for scaled statistical observations, replaced the tradition of recording numbers directly onto the map. Statistical maps commanded widespread respect as a vehicle for demonstrating causal connections between spatially correlated phenomena. As conventionally told, it was the medico-statistical mapping of cholera patterns in mid-nineteenth-century European cities, especially Edwin Chadwick's and John Snow's 1840s maps, that secured the scientific status of statistical mapping (Fig. 9.3). But as Tom Koch's historical analysis of this story insists, disease maps acted as propositions rather than scientific representations and their scientific status should be regarded as such.[23] Nonetheless, while the topographic map represents the specificity and uniqueness of

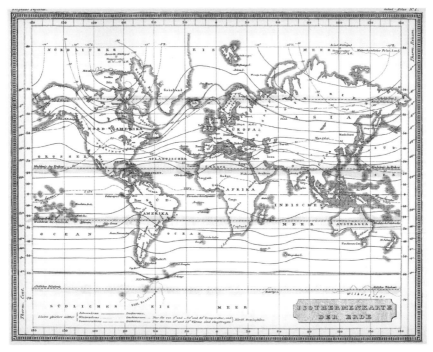

Figure 9.2 Heinrich Berghaus, world map showing isotherms, based on
Alexander von Humboldt, 1845

places, the thematic map is a tool for nomothetic thinking and thus for
social planning. Dot maps, for example, were especially popular among
medical doctors committed to the neo-Hippocratic thesis of environmental
causes for disease, until Pasteur's work refocused attention on internal and
biological causes.

Thematic mapping and science

The authority of thematic mapping derives from the statistical foundations
of the information it conveys. Given that such maps are generally produced
by agencies for the better management of local territories, the problems of
transporting knowledge are often not as great as they are with exploratory
geographic or topographic mapping. But thematic mapping suffers from
two fundamental and often unacknowledged weaknesses, beyond the obvious methodological and design problems of interval and scaling choices.
First, the spatial correlations which such maps suggest can too readily be
interpreted as causal: what is called the 'ecological fallacy'. An example of

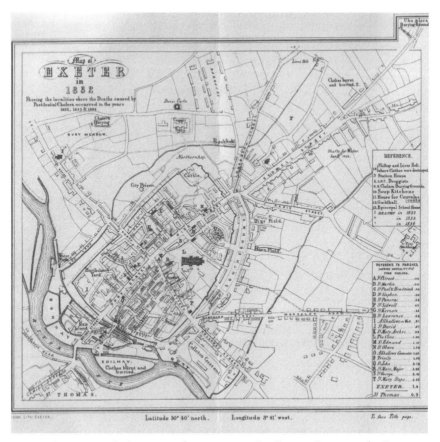

Figure 9.3 Cholera map of Exeter, 1832 (Bodleian Library, Oxford)

this is the nineteenth-century medical hypothesis that high olive oil consumption might be the cause of hiatus hernia, a belief based on an assumed spatial correlation between the incidence of the disease and olive oil within the Midi diet. It was Malgaigne's 1840 *Carte de la France hernieuse* ('Hernia map of France') that undermined the claim, illustrating the incidence of hernias among the population of French departments by shading at six intervals the numbers of people per 'hernious' individual within each administrative unit, and then superimposing dietary boundaries to reveal possible causes of the variation.[24]

The second weakness of the choropleth map as a scientific tool is iconographic: the choice of colour shading to illustrate interval differences. The concept of 'shading' itself carries powerful moral connotations; this was especially true in an era of self-conscious 'enlightenment', when darkness and shadow implied ignorance and decay, both physical and moral. Thus

Edward Quinn's 1830 *Historical Atlas* used a similar device to Abraham Ortelius's parting clouds to surround his images of the known world at different stages in human history. But in Quinn's case they reveal the darkness of ignorance being pushed progressively aside by the onward march of civilization through the ages. On nineteenth-century thematic maps, colour shading consistently used dark tones to register failure in the spatial narrative of progress. Thus French statistical cartography throughout the century, in maps of educational standards or poverty and social provision, regularly divided the country into two parts: an 'enlightened' north where lighter shades dominated, and an 'obscure' France south of a line from San Malo to Geneva, whose failures were dimly visible through the cartographic gloom. A similar use of shading and the ecological fallacy is to be found on what today appear extraordinary choropleth maps published in scientific geographical journals to illustrate correlations between the distribution of 'climatic energy' and that of 'civilization' (Fig. 9.4).

If statistical mapping lost favour among medical researchers in the later nineteenth century, its popularity among social planners and 'hygienists' peaked in the early years of the twentieth century, as the 'civilization' maps suggest. Statistical survey and mapping of what were deemed cultural traits, such as language, dialect and custom, were central to the arguments of supporters of *volk* and *nation* in determining the territorial

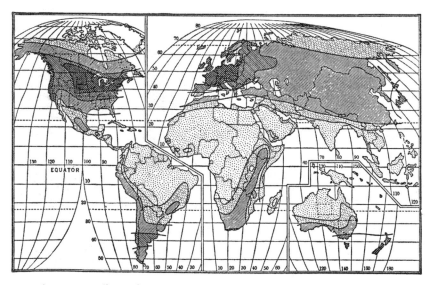

Figure 9.4 Ellsworth Huntington, World isoline map showing levels of 'civilization' (*Mainsprings of Civilization*, 1945; author's copy)

extent of the unifying German Reich in the 1850s and 1860s.[25] From the work of social statisticians and statistical cartographers such as Frédéric Le Play in France, Henry Mayhew in Britain, and Heinrich Berghaus in Germany, and American geological and soil surveyors such as John Wesley Powell or Eugene W. Hilgard, emerged the idea of mapping as a critical tool of state policy in the 'Progressive Era' at the turn of the twentieth century. The concept of the survey was transferred from the exploration and inventory of colonized lands to hidden features of metropolitan societies. In the minds of social philosophers such as Patrick Geddes, a unified survey of the physical and social characteristics of a region could provide an ecological portrait of community in place, a foundation for its rational planning, and a stimulus to the civic virtue and community participation of its people.

The survey did not require exploration of unknown spaces, but an 'archaeology' of interconnected relations within a place that could be revealed by the very process of mapping. Such mapping would most appropriately be undertaken by the community itself, working under instruction, and in the process contributing to the further consolidation of civic virtues. Public display of the resulting maps would further this goal. Thus, in the United States, various states and counties in the 1920s, especially in New England and the Mid-West, encouraged schoolchildren to participate in local surveys that would reveal the true state of their community, especially regarding its physical and moral heath. These maps had the effect of moral self-regulation, as in the example of a publicly displayed Springfield, Massachusetts survey map that flashed 1250 coloured lights indicating the distribution of babies born in 1913, distinguishing between green lights for homes where the birth was registered, and red for unregistered births (presumably out of wedlock). In Britain, the survey movement peaked in the 1930s with the recruitment of schoolchildren from across England and Wales to map the use of every parcel of land, returning the results to London where they were compiled onto topographic base maps and coloured to reveal national patterns of land use. Deep reds and purples dramatically revealed the octopus sprawl of great cities such as London whose suburban anonymity was supposedly threatening traditional civic virtues[26] (see Plate 6).

The authority of survey mapping peaked in the post-war years of welfare-state planning and modernist social engineering, which depended heavily on map overlay techniques both to analyse social patterns and 'problems' and to develop persuasive arguments for their solution. The 'plan', in which the map played as significant a role as the written text, became a central instrument of policy. Even as comprehensive spatial planning came under serious criticism in the late 1960s and early 1970s, the value of the cartographic techniques upon which it depended was strongly reasserted in ecological landscape design. The influential Scottish writer Ian McHarg recommended the use of polygonal overlays to connect various landscape features:

> by starting with bedrock geology and then surficial geology, and then reinterpreting these to reveal groundwater hydrology, you explain physiography and also surficial hydrology. This then leads you to inevitably to soils, which leads you to plants, which leads you to animals, which can lead you to land use. . . . every one of these steps is in fact either correcting or reinforcing.[27]

McHarg is commonly credited with pioneering the cartographic techniques now computerized within Geographical Information Science. GIS draws upon and synthesizes a range of spatial representations – the aerial photograph, remote sensed image and topographic map – as well as thematic spatial statistics to generate complex overlying patterns, enhanced and manipulated by the computer. The resulting images have enormous graphic power, reshaping our vision and understanding of the world and our capacity to intervene in its material and social processes. Combined with the capacity to transmit data instantaneously and manipulate it in real time on the PC screen, we have hugely increased the cartographic illusion of synoptic vision and action at a distance (the magic of maps). But we need to remain as alert as ever to the persuasive power of these newer cartographic images. The widely reproduced NASA image of 'Spaceship Earth' that I have referred to already in these essays, and whose authority stems from its photographic 'realism', has been read universally but uncritically as a sign of a vulnerable globe threatened by anthropogenic environmental crisis, and thus as a moral mapping of human treatment of a nurturing mother earth. In fact the image itself contains no evidence that necessarily supports such a claim. The Apollo photograph, like the nineteenth-century disease map or a twenty-first-century GIS image, is propositional as much as it is representational.[28]

Mapping and circulating knowledge

The story of how the whole-earth image was obtained by the Apollo astronauts, returned to earth, circulated globally, interpreted and gained agency in the discourses of environmentalism and one-world globalization, exemplifies many of the continuities in mapping as a spatial process.[29] In his work on the making and circulation of scientific knowledge, Bruno Latour has used the term 'immutable mobile' to characterize those material agents that permit scientific discourse to sustain its claims of empirical warranty and repeatable truth in the absence of eyewitness evidence.[30] The map is a perfect exemplar of the immutable mobile: a container of information gathered at specific locations, returned to a 'centre of calculation', and then placed once more into circulation as a vehicle and instrument of scientific knowledge and further hypotheses. The entire history of cartography can be told as a history of struggle to realize such a status for the map. Thus Claudius Ptolemy, whose tables of locational coordinates introduced modern mapping to Renaissance Europe, may never have drawn actual maps. His book gave sufficient information for a skilled reader to construct a projection and plot the coordinates necessary to produce the maps from his data. Text and tabulated figures are far more easily and accurately copied and transported than a set of drawn maps. Securing the immutability of the mobile has been a constant obsession of cartography. It is fundamental to the map's claim to be more than an imaginative picture. Indeed, cartographers have actually drawn upon the authority of cartographic procedure to grant legitimacy to what were in fact complete fabrications. Thus the sixteenth-century French cosmographer André Thevet plotted lines of latitude and longitude around maps of completely illusory islands.[31] Such charlatanry reveals the ultimate impossibility of the cartographic conceit. The only true map is the territory itself, as Louis Borges long ago pointed out.

The search to secure immutable mobility for the map reveals another feature: cartography's prosthetic quality. The map is one of those instruments that serves to extend the capacities of the human body. Like the telescope or microscope, it allows us to see at scales impossible for the naked eye to see and without moving the physical body over space. The thematic map reveals the presence of phenomena that are beyond our normal bodily senses, as for example a trend surface map of property values or of air pollution. The map also has a powerful recursive quality, at once a memory device and a foundation for projective action. This is immediately apparent in European mapping in the 'age of discovery', where the map was at once a necessary starting point for the exploration process and a principal outcome of that process.

These prosthetic and circulatory aspects of mapping are true also for the social survey, and they remain so for the most technically advanced mappings of today. It is these features of the mapping process that makes it such a fertile and powerful epistemology in knowing and representing the world. The map is at once empirically rooted and imaginatively liberated and liberating. No spaces can be controlled, inhabited or represented completely. But the map permits the illusion of such possibilities. Mapping is a creative process of inserting our humanity into the world and seizing the world for ourselves. This is why today the boundaries between the art and science of mapping, so long and so arbitrarily surveyed, charted and policed, are increasingly smudged and faded, and why the imaginative and projective potential of mappings has become so vitally present in contemporary life.

10 Carto-city

Urban space and cartographic space are intimately related. This is so historically, practically and conceptually. Urban origins, at Ur or Sumer for example, are revealed through the mapped reconstruction of their street geometry and building plot plans.[1] Buildings eventually fall, but ground plans remain; elevation is the least durable element of urban form, but plot forms its longest lasting element. Urban archaeology reconstructs urban places through peeling back successive surfaces, like the pages of a historical atlas, and mapping the stratigraphy of material deposits stretched across former urban space. From such material mappings we reconstruct not only a city's physical appearance but also its social, political, commercial and religious life.

Practically, a visitor's confrontation with an unfamiliar city is typically mediated by the map: of transit routes, of streets, of tourist destinations. Urban experience in a new city is often a process of negotiating the divergence between cartographic and material spaces. In cities with complex underground rail systems, such as Tokyo, London or Paris, it can take years before the point pattern of stations familiar from the subway map is fully coordinated with the surface experience of townscape. In great metropolises, possession and use of a standard street directory such as London's pocket A–Z guide or Los Angeles' *Thomas Guide* are almost signs of citizenship. Appropriately to their different densities, London's *Geographers' A–Z* guide can be held in the hand and carried in the pocket. (*Fig. 10.1*). Los Angeles' *Thomas Guide*, by contrast, is designed to sit on the passenger seat of a car. These handbooks have become cartographic icons of the cities whose streets and addresses they designate. Much more than functional instruments, aids to fixing destinations or following routes, they are bearers of urban meaning and character: the map becomes to some degree the territory.

Conceptually, the map has often either preceded the physical presence of the city or served to regulate and coordinate its continued existence. St

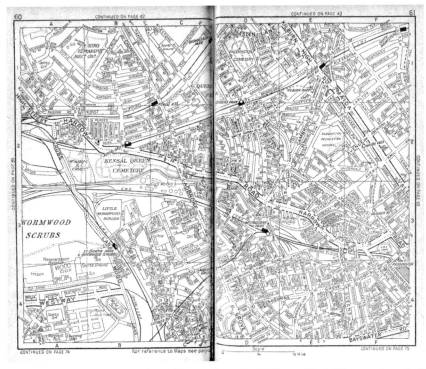

Figure 10.1 Kilburn area of London, *London A–Z*, 1968, pp.60–1 (Geographers A–Z Company, London)

Petersburg, Washington DC, New Delhi, Brasilia, and countless fortress and colonial cities from ancient Greece onwards, existed on paper (or its equivalent) before they had any material expression. Paris, Rome, Vienna, Amsterdam, Beijing, Jerusalem – virtually every great city – has been either reconstructed or expanded by means of a drawn plan.[2] And beyond the physical extension or reconstruction of urban space, the map has both recorded and determined countless aspects of urban life and citizenship. Maps of disease and morbidity, for example the cholera cartography of Victorian London or Philadelphia, discussed in the previous chapter, helped make the modern metropolis a survivable space in the face of those viruses and bacteria that thrive on human density.[3] Maps of social and ethnic status have shaped the political life of urban democracies, nowhere more dramatically than in the case of twentieth-century American zoning maps, used by housing and loan companies for 'red-lining' inner-city ghettos and later by government agencies to assert and monitor civil rights. In every way, the map registers the city as a distinct place and a unique landscape. Cartography acts not merely to record the various ways that the city is materially present, but as a creative intervention in urban space, shaping both the physical city and the urban life experienced and performed there.[4]

The ubiquity of cartography as a dimension of urban life and form makes a comprehensive survey of their relations impossible. I focus here on ways that the urban map is positioned between creating and recording the city. It is this dual function that releases the imaginative energy of mapping, and which has consistently attracted the attention of artists as well as technicians to urban mapping. From the vast archive of urban maps, plans and artistic interventions into urban mapping, I explore how the modern city as material and social space interacts with the map as scientific instrument and artistic representation of its space and life.

The complexity of the interaction is dramatically apparent in a post-modern American city such as Los Angeles, Houston or Phoenix. These are among the most intensively mapped spaces in the history and geography of the planet: every square metre is geo-coded by government and private or commercial agencies for purposes ranging from environmental protection, public health and safety, efficient transportation and taxation to property insurance, marketing, political persuasion and religious proselytizing.[5] Maps have played a critical role in shaping their physical spaces and land uses, and continue to control the daily lives of citizens through zoning ordinances, ZIP codes and the myriad territorial regulations that shape urban daily life. Theoretically, scientific cartography should make these cities highly rational, coherent spaces. Indeed, flying over a city such as Houston, a cartographic order is immediately visible in the repetitive grid of major streets, and within this in the curvilinear geometry of residential roads and house lots, or in the distribution along arteries of office, production, stor-age/distribution and retail spaces whose rectangularity reproduces the grid at a larger scale. Yet, on the ground, such cities are among the least 'legible' places on earth. Moving across their surfaces, an individual familiar with the spatial sorting of land uses and the hierarchy of distinctive symbolic struc-tures that characterize the urban structure of earlier cities is confused, even assaulted, by the seemingly random scatter, individuality of design and consequent visual chaos of urban elements. The twenty-first-century urban landscape seems to confirm the problem of legibility in the constant and competitive presence within it of words, phrases and whole texts: bill-boards, street signs and posted ordinances.[6] The volume of written language in the public spaces of contemporary cities far exceeds that in more trad-itional ones. Image and text, whose effective harmonizing is cartography's signal contribution to spatial representation, have become disjointed, and their falling apart denotes the erosion of a relationship that underpinned urban modernity.[7]

This does not signify a reduction in the map's role in urban place-making and experience. Indeed, a characteristic way of negotiating movement within the post-modern American city is the computer-generated map, custom produced for any destination address, from any point of departure within the United States. Map Quest© can create an instant digital image of any urban location at any requisite scale, using a simplified set of standard colours and cartographic signs. Such images are resolutely functional, entirely ignoring the context of the places they represent, utterly unconcerned with the urban iconography that traditionally expressed *civitas* or the city as public space. No urban map could be further removed from an image such as Jacopo de' Barbari's map of Venice in 1500 (Fig. 10.2). The paradigm example of early modern city mapping, de' Barbari's detailed

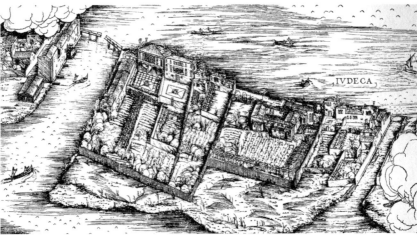

Figure 10.2 Jacopo de' Barbari, *Venetia*, woodcut map, 1500; the detail (below) shows the Giudecca area (Museo Correr, Venice)

172

townscape guided the observer not to a physical destination but to the civic spirit of Europe's greatest commercial city.[8]

'Scientific' cartography's inability to capture the apparent incoherence of the contemporary city has, however, opened new possibilities for urban mapping. As I shall illustrate, these have been colonized creatively for a wide variety of projects: conceptual, political, and purely ludic, breathing new life into the connections between city space, city life and mapping.

Radial axis and the grid

Geometry, specifically the radial axis and the grid, underpinned both scientific cartography and modern urban form. Their power and historical endurance in both the map and the city lies in their combination of practical and symbolic efficacy. The circle's 360 degrees generate a 'centre enhancing' axial form focused on a single point. Functionally and symbolically, this extends power panoptically to the horizon, encompassing a potentially infinite territory. Versailles, Karlsruhe, baroque Rome and Second Empire Paris have all inscribed this simple geometry into urban space. The same axial pattern emerges from the simplest mapping technique: taking radial sightlines from a single point. Multiple survey points can be selected randomly along a pathway or be connected to a cardinal base line. Back-sighting confirms positional accuracy, producing a network of intersecting axes, such as the wind-related rhumb lines on a marine chart. The magnetic compass connects these sightings to global and even celestial geometry. The radial planning of a city is at once practical and symbolic, as the Roman architectural writer Vitruvius recognized in his description of the city perfectly oriented in respect of wind directions. The first recorded systematic urban mapping was the 'description' of Rome undertaken by Vitruvius's first modern popularizer, Leon Battista Alberti. No actual map exists for Alberti's project, merely a set of coordinates produced for known locations along axes converging on a single point of observation, and this may indeed be as much as Alberti intended to produce.[9]

The alternative geometrical form shared by urban plan and mapping is the grid or chequerboard of orthogonal lines crossing at right angles. While radial axes enhance the centre, the grid is 'space equalizing', infinitely extendable over the surface and privileging no single point, but rather reducing each to a unique coordinate. The grid generates the simplest and most widespread form of urban plan. It is found in the earliest Greek colonies in the Mediterranean, in the design of imperial Chinese cities, in Spanish New World pueblos, in the Tudor planted towns of Ulster and the urban settlements of the USA. The colonial urban form *par excellence*, the grid can also be an expression of urban democracy, equalizing lot sizes and maximizing the ease of platting and disposing of urban land into private property. As William Penn's Philadelphia design reveals, the grid can easily be elaborated to generate open spaces – squares – to interrupt its monotony and produce public space within an otherwise privatized territory.

The grid performs a similar function in mapping. Introduced into modern Western cartography in the early fifteenth century through Claudius Ptolemy's *Geography*, but known to Chinese cartographers at least 300 years earlier, the introduction of a graticule of longitude and latitude lines as the basis for determining location and translating the sphere into a two-dimensional map is by far the most significant feature of modern cartography.[10] The grid is a ubiquitous location-fixing device. Unrelated to planetary coordinates, the grid can be stretched across any spatial scale, as the British Ordnance Survey's national grid or the lettered and numbered squares superimposed on countless urban maps demonstrate.

These shared geometries of urban design and cartography are instrumentally effective and symbolically significant. They are conceptually easy to grasp and relatively simple to use in spatial design and representation. Neither is restricted to a single scale but may be applied from microcosm to macrocosm. Thus, each not only possesses specific symbolic attributes – power and panopticism, reason and democracy respectively – but also shares the capacity to connect mundane space to the cosmic patterns, movements and logic. They therefore propose and permit urban mapping as a philosophical exercise.

The philosophy and ethics of the urban map are apparent across the history of modern urbanism. The profound impact of Vitruvius's urban designs in Renaissance Europe derived from their appeal to humanists and natural philosophers engaged in rethinking both the nature of urban life and the mapping of global space. The 'ideal city' debate, conducted among architectural writers from Alberti and Francesco di Giorgio Martini to Sebastiano Serlio and Vincenzo Scamozzi, concerned more than the formal design of urban space. It was about mapping urban life and citizenship.[11] Thus the city represented in the anonymous urban image now in Baltimore (see *Fig.* 5.3) represents much more than a symmetrical grid of buildings, streets and open spaces rendered in deep perspective. It also maps an image of citizenship, derived from Republican Rome and the Stoic writings of Cicero, Seneca and Marcus Aurelius. The cardinal virtues of Justice, Prudence, Temperance and Fortitude stand atop the four columns that define its central square. The buildings gathered around that space correspond to the public functions that regulate urban life. And across the foreground of the image creeps the bent and burdened, but immensely dignified, figure of the Stoic. Acknowledging the cosmic order mapped into urban space, and subordinating body to mind, he signifies the good citizen's duties of reason, reverence and sociability. This map of urban space is also the map of urban virtue.

Axial and grid geometries were united in the service of modernity's grandest social project: the creation of the USA, as represented in the design of America's federal capital. Positioned within the cardinal square of the federal district, Charles L'Enfant's map for Washington DC combined the democratic principles of the Constitution and Bill of Rights in the chequerboard of residential streets, with the inscription of federal authority in axes radiating from the separate seats of executive and legislative power. The plan generated 15 public open spaces, one for each state of the Union at the time of the city's foundation. 'Universal' principles embodied in the new republic were thus mapped into its capital.[12]

Geometry, the map and urban legibility

Early modern and Enlightenment city planning saw in geometry a medium of legibility. The city was to be read as a text: by and for its rulers, its citizens and its visitors. Printed urban maps expressed and reinforced the city's legibility. These were not initially intended primarily as location markers or way-finding instruments, except for a small number of guide-book maps such as those describing the monuments and pilgrimage sites of Rome. The earliest urban maps are overwhelmingly celebratory, intended to frame in a comprehensive image the city's complex social and spatial total-ity. By the late sixteenth century such maps formed a distinct genre, having emerged first in Europe's most heavily urbanized, mercantile regions: upper Italy, southern Germany and Flanders. Some Italian urban maps such as Leonardo da Vinci's map of Imola adopted an orthographic perspective; others took the high angle, bird's eye view apparent in de' Barbari's Venice. Northern maps were more commonly townscape views, graphing the urban skyline in silhouette along a horizontal perspective.[13] While the former universalized the city by representing it according to a standard set of geometrical rules and surveying principles that reflected universal ideals of order, the latter particularized the city through the unique elements of its skyline: cathedral, guildhall, parish churches. By the middle of the century that distinction was fading. Nicholas Crane captures the impact of changes in urban mapping by describing the cartographer Mercator's reaction to Hieronymous Cock's 1557 view of Antwerp:

As long as Mercator could remember, Antwerp had been pictured from the water, as a bustling, dishevelled river port. Cock's viewpoint was high above the opposite side of the city. Antwerp had become a disciplined urban network of streets and civic symbols surrounded by massive geometric defences. Wagons entered landward gates, bound for fleets of patient ships on a distant, placid Schelde. Cock's Antwerp was a celebration of mercantile might, and a working diagram of a modern city.[14]

The paradigm of early modern urban mapping projects was the multi-volume urban atlas, *Civitates Orbis Terrarum*, published in Cologne from 1572 and edited by Georg Braun and Franz Hogenberg. Intended to illustrate every major city in the world according to a standard, printed cartographic format, it was conceived in response to Abraham Ortelius's world atlas of 1570. Earlier encyclopaedias and cosmographies had illustrated cities, but while some places were well served by recognizable images – Constantinople, Rome, Paris and Venice for example – many of the named cities in Hartmann Schedel's *Chronicle* of 1493 or Sebastian Münster's *Cosmographia* of 1544 had been illustrated by generic townscape images. Braun and Hogenberg gathered printed city maps from local sources to ensure the most accurate and up-to-date image of specific cities, allowing the atlas owner to survey the *civilized* globe of urban places within the privacy of a study or reading room. While some images adopted the urban silhouette, the favoured perspective was the bird's eye view, exemplified perfectly in the maps of Cologne or Amsterdam. The city is revealed as a theatre, seen from an elevated point far above and beyond its confines, at an angle sufficient to reveal both its plan pattern of streets, squares and open spaces, and the elevation of its principal buildings and monuments. Distant, to be sure, yet close enough for the rhythm of its life to be pictured in the pedestrians, carriages, wagons and ships moving on its roads and waterways, the city is immediately legible as a coherent community. Its citizenry and their distinctive customs are denoted in the costumed figures who regard the map user from within the cartographic space declaring their citizenship as the principal determinant of their identity.[15]

Decorating their maps with the city's coat of arms and those of its great families or principal guilds reinforced Braun and Hogenberg's emphasis on unity and civic order. Cartouches and printed text press the point further: each city is honoured by the depth of its history, the nobility of its citizens, the wealth of its merchants and the beauty of its buildings. The urban map is synthetic rather than analytic; its goal celebration rather than analysis or critique. This is true even of the two New World city maps included in the collection: the Aztec capital Tenochtitlan (see *Fig.* 3.5) and the Inca city of Cuzco. As discussed in Chapter 3, the former map, based on Cortez's own illustration of the Aztec capital, illustrates the very antithesis of *civilization* in the classical sense of civic virtue lived out in urban space. Its central square is dominated by the great ziggurat upon which human sacrifices are being performed. Cartographic parallels between the island cities of Tenochtitlan and Venice reinforced the Aztec capital as 'Other' to Europe's self-proclaimed model of civic perfection, yet Braun and Hogenberg's urban mapping principles subordinate such difference to stylistic consistency.[16]

Legibility, improvement and control

The decorative, celebratory style of urban mapping exemplified by *Civitates Orbis Terrarum* dominated European urban mapping into the eighteenth century. Urban plans projecting newly founded cities or urban expansion adopted a more severe, undecorated style as part of their rhetoric of practicality. This is already apparent in Vincenzo Scamozzi's 1599 plan of the Venetian fortress city of Palmanova, in John Evelyn's 1666 plan for rebuilding fire-damaged London, or in the 1703 drawings for St Petersburg. Their draughtsmanship provided the model for the 'scientific' urban maps of the Enlightenment such as Michel Etienne Turgot's 1739 *Plan de Paris*. Urban legibility becomes the overarching goal of Enlightenment city mapping, to be achieved through precisely measured survey using carefully calibrated instruments. The undecorated simplicity of eighteenth-century graphic design articulates goals of cartographic accuracy and objectivity by erasing evidence of human intervention between survey instrument and printed image.[17] The severely constrained outline of carefully surveyed streets, open spaces, building footprints and plots, the absence of citizens in distinctive costume and of cartouches and inserts narrating the town's history, together with monochrome printing from fine-line copper engraving or later lithography, all signify new ways of thinking about the city and about urban life. The traditional vision of the city as a self-governing *polis* or Christian community had underpinned earlier cartographic emphasis on civic harmony, identity, community and dignity within a unified urban space. This vision was being eroded by the city's subordination within the emerging territoriality of the nation-state, along with modern secularism and individualism, and by urban population growth and spatial expansion, industrial production, and new forms of social cleavage and solidarity. In response, the city was reconceived, and new ways of imagining and experiencing urban life were reflected in maps whose intent was increasingly analytic rather than synthetic.

By the mid-nineteenth century, surveyed urban plans had become the base maps for the emerging science of urban statistics, by means of which expanding state capitals and new industrial cities were to be regulated. As we saw in the previous chapter, cholera and typhoid, poverty and prostitution, alcohol consumption and criminal 'deviance' – all regarded as primarily urban ills – came to be understood through the medium of the urban map.[18] The accuracy and authority of the base map was fundamental to the persuasive power of the statistical information plotted onto it. Rather than celebrating the unity and harmony of urban community, the map's task was to bring into the light of practical reason invisible but all-too-potent urban pathologies. Their amelioration often itself involved further mapping of urban space: clearing and re-planning 'crowded districts,' laying water supply and drainage systems, platting new suburbs, cemeteries and parklands.

The celebratory aspect of the urban map did not entirely disappear. In the United States, for example, it gained a new lease of life in mid-nineteenth-century county atlases lauding western expansion. The bird's eye perspective was perfect for demonstrating the elegance and prosperity of newly established and often scarcely built towns.[19] Emphasizing their grid of streets and the bustle of carriages and carts, the maps promoted what were often in reality chaotic, unregulated places as well-ordered civic communities. They contrast strongly with the Sandford Company's fire insurance maps from the same years whose functional goal of assessing risk and determining premiums is apparent in the severity of unembellished detail, coding such data as street width, building height and constructional materials, or the proximity of fire hydrants. Anticipating the zoning and planning maps, insurance cartography, more than the celebratory urban images, foreshadowed the dominant direction of urban mapping in the succeeding decades.

Controlling metropolitan cities was a dominant theme of twentieth-century urban mapping. The response to 'monster urbanism' is recorded in maps, with the modern metropolis constantly threatening to outstrip the map's capacity either to make it legible or to regulate its material and social disorder. London, the first world city of the twentieth century, is a prime example. The seemingly uncoordinated and uncontrollable sprawl of its suburbs, dramatically accelerated by mass transportation and later by car ownership, anticipated the forms and processes that now dominate city landscape and life globally. One heroic, individual response to metropolitan legibility was Phyllis Pearsall's 1935 creation of The London A–Z, a pocket atlas with gazetteer of every street address in London county and the suburbs beyond (see Fig. 10.1). Her hugely successful, commercial project began in response to a personal dilemma, that of finding an individual address without an adequate street guide to London. To create her urban atlas she walked 23,000 streets and a distance of nearly 5,000 kilometers.[20] The outcome was a work that remains a handbook for every Londoner, but it reduces every urban element to the same format, abandoning any semblance of the city as a coherent urban structure in favour of its legibility as a continuously coded surface.

A more bureaucratic cartographic attempt to control the British metropolis was the 1932–4 National Land Use Survey in which schoolchildren were deployed as part of their geographical training for citizenship to record land uses across England and Wales, including 'non-productive' urban land. As we have seen, the London plate of the resulting national map series uses a livid purple to illustrate vascular urban tentacles strangling the soft green of rural England.[21] It can be read alongside the 1904 plan for Letchworth garden city or the 1969 street plan of Milton Keynes which mapped alternative urban visions, designed to control London's spread (see Plate 6). These twentieth-century cartographies represent the apotheosis of the map as an instrument of urban policy, not only to recapture the legibility of the modern metropolis on paper, but to sustain its physical and social coherence as a material space.

In their own ways, such city maps reveal an unsustainable Modernist faith in geometry as the guarantor of urban legibility. Conventional Euclidean principles and forms neither describe nor contain the spaces of the increasingly flexible, mobile, cybernetic city. Since the late eighteenth century, the free-flowing serpentine line has battled with orthogonal geometry as the privileged design medium for expressing the triumph of individual, 'natural' man over the classical or Christian model of the citizen.[22] The serpentine line, apparent in John Nash's work in London's West End connecting St James's and Regent's parks and present as a relatively minor element in the picturesque Letchworth plan, is omnipresent by the mid-twentieth century in the design of Milton Keynes. It dominates post-modern urbanism in both street plan and built form. Indeed, high-tech mapping techniques help assure its triumph. Computer-based digital laser technologies allow twenty-first-century architects to deform conventional building structures in response to new materials.[23] These are mapping techniques that will soon be applied beyond such signature urban buildings as Frank Gehry's Guggenheim Museum in Bilbao or Los Angeles' Disney Concert Hall, to the form of the city as a whole. Scientific mapping remains more successful in projecting the future form of the city than in capturing the legibility of its daily life.

New legibilities: the artist, the map, and the city

Pearsall's London A–Z project was precisely intended to make London legible for everyday life. We might contrast it with an entirely different but contemporary mapping of the modern metropolis, that of the German cartographer, Hermann Böllmann. Armed with a technique known to nineteenth-century artist-cartographers as *Vogelschaukarten*, and which dates back at least to Jacopo de' Barbari's Venice image, Böllmann confronted modernity's most demanding urban landscape: Manhattan. Using 67,000 photographs, 17,000 taken from the air, he created in 1948 a hand-drawn map image that captures precisely the soaring quality of the New York skyline while rendering street plan and building plots with remarkable accuracy and clarity. Pearsall's and Böllmann's distinctive mappings of the mid-twentieth-century metropolis help to illustrate a debate over how urban space is known and experienced and thus how it should be mapped (Fig. 10.3).

Like Pearsall's, Böllmann's concerns in making the city legible through cartography were principally commercial. But in the immediately succeeding decade, the idea of urban legibility became a dominant concern of progressive urban thinkers on both sides of the Atlantic. In the late 1950s the American urban designer Kevin Lynch drew upon the economist and polymath Kenneth Boulding's highly influential text *The Image* to design a research project that used 'mental maps', drawn by ordinary individuals interviewed about their knowledge of the urban spaces in which they lived and worked, to make general statements about 'urban legibility', that is, the clarity of urban space in everyday life.[24] Cities with strongly defined formal elements – simple geometry of plan, clearly defined district boundaries (edges), landmarks and other easily visible features – such as Boston, were defined as more 'legible' than less formally or more loosely structured urban spaces such as Newark, NJ and Los Angeles respectively. For Lynch and his many followers the citizen's ability to map the city became the principal measure of urban environmental quality, and strongly visible,

Figure 10.3 Hermann Böllmann, *New York, a picture map*, 1962. Detail of Manhattan showing Battery Park area (Pictorial maps Inc., UCLA Library)

material features of plan and elevation were promoted as determinants of successful urban design. The eye was figured as the privileged organ of urban experience. Cartography and mapping are thus strongly privileged as the media through which the city is known and improved.

In the same years, a more critical and politicized perspective on urban mapping was adopted by a group of French political radicals whose ideas had originated in avant-garde artistic movements associated with Surrealism. The Situationists, led by Guy Debord and inspired in part by Walter Benjamin's celebrations of urban space and life, celebrated the *dérive* or 'drift' as a way of experiencing everyday life in the city free from the attempts of authority to plan and regulate urban movement.[25] They explicitly associated the panoptical map with rational, alienating modernity. Pearsall's London rather than Böllmann's Manhattan might be taken as the paradigm mapping mode for this way of relating to the city, which derived from walking and taking the perspective from the street. Situationist urban mapping sought to emphasizes the breadth of embodied, sensory experience in the city, refusing all hierarchy of urban locations and celebrating the casual and adventitious aspects of urban life. The Situationist critique of the totalizing urban plan in favour of the city as performance art has been adopted by later urban progressives such as Michel de Certeau and Rosalyn Deutsche, opening the idea of urban mapping to a range of artistic intervention over the closing years of the twentieth and the opening years of the twenty-first century.[26] Their 'mappings' may deploy the analytic capacities of scientific cartography, often using advanced technologies to rework some of the goals of earlier urban mapping. They seek to capture legibility from the contemporary city, not as a means of reworking its material spaces, but as a way of enhancing the experience of everyday urban life.

9/11: urban mapping and art

Space permits discussion of only one example of such creative urban mapping. But it is a powerfully telling example in its symbolic and ethical dimensions. De Certeau's celebrated critique of scientific mapping's distanced and totalizing vision of the city in *The Practice of Everyday Life* took as its model the view of Manhattan from the top of the World Trade Center, the very perspective that Böllmann's map celebrates. Ironically, the destruction of those towers on 11 September 2001 presented perhaps history's greatest single challenge to urban mapping. Maps and plans of every urban system affected by the attack – transportation, utilities, communications, air quality – and new cartographies detailing the rapidly evolving impacts of the buildings' collapse on all aspects of city life, were vital to the response mounted by New York's myriad public and private agencies. Cartographers used the latest Geographic Information Science (GIS) technology to coordinate and plot these diverse data sets in real time. Maps could not be carefully redrawn in such a rapidly and continuously evolving situation, as information fed in from scanners, satellites and photogrammetric surveys had to be integrated with existing maps and data bases for the affected zone. Over a three-month period more than 2600 maps were produced, using 'techniques of layering, seriality and transparency, complemented by the destabilizing power of interactivity, movement and animation'.[27] Many were made immediately available on the Internet.

Over the extended period of recovery, interest among the public in the site of the attack and collapse remained intense: victims' relatives, New York citizens, distinguished visitors and ordinary Americans came to bear witness and observe progress at Ground Zero. The scale of the urban footprint and the barriers erected to protect the recovery work made it almost impossible for casual visitors to comprehend what they saw. In response, the artist Laura Kurgan, who had previously used mapping techniques to explore a range of issues surrounding political violence, gained funding for the production of two editions of a map: *Around Ground Zero*, published in December 2001 and March 2002. Her maps use primary colours and simple graphics to identify key elements of the site: viewing platforms, temporary memorials, cranes and trucks as well as variously demolished or damaged buildings. The project had to negotiate the most delicate of ethical dilemmas, given the implications of viewing a scene of mass murder from which human remains were being actively removed. They were distributed free to visitors.[28]

The contemporary city presents both complex new challenges and enormous opportunities for mapping, as do emerging survey and plotting technologies. Indeed, the map may be the only medium through which contemporary urbanism can achieve any sort of visual coherence. There remains a strong, if unrecognized, celebratory dimension to urban mapping, not merely in the banal sense of cities' self-promotion through advertising or tourist maps and plans, but in the choice of the scale, content, design and colour of the myriad cartographic devices (many today interactive) developed by public agencies and private bodies to communicate and regulate contemporary urban systems and processes. The goal of rendering legible the complex, dynamic and living entity that is a city remains an urgent one. But today's acute awareness that cartographic images can never be innocent vehicles of information dissolves neat distinctions between celebratory and regulatory urban maps. Urban space and cartographic space remain inseparable. As each is transformed the relationship between them alters, and current visual technologies mean that the opportunity for creativity in shaping and recording urban experience is greater than ever, as too is the need for critical attention to the making and meaning of both public and private urban spaces.

VI Metageographic visions

11 Seeing the Pacific

On 13 March 1944, readers opening the *Los Angeles Times* at page 3 would have faced a full-page coloured map showing the struggle for Wake Island, then being contested between USA and Japanese forces in the Western Pacific (see *Plate* 7) A dramatic bird's eye view of the island and its harbour centres a representational space extending to a hemispheric horizon. An air and naval battle dramatically frames the curving horizon. The strategic significance of the battle for Wake within the global struggle between the USA and Japan to dominate the Pacific is explained to the reader by means of text inserts, together with directional and iconic devices such as arrows and aeroplanes to explain the tactics of the struggle. A month later, on 17 April 1944, the *Times* carried a similar pictorial map illustrating the vast scope of American air and sea operations in the Pacific theatre. Against a globular representation of the central Pacific Ocean, an outline of the continental USA at the same scale indicates the area and distances involved in the conflict. Here, too, a curving horizon frames dramatic battle scenes, together with an image of the White House in Washington DC from where the war is being directed. The *Times* staff artist, Charles Hamilton Owens (1880–1958), drew both maps, among nearly 200 similar images printed in the newspaper between early 1942 and the war's close in 1945. Today forgotten, but in his lifetime justly acclaimed within and beyond Los Angeles for the quality, accuracy and drama of his illustrations, Owens was one of a number of artist-cartographers who responded with highly innovative and imaginative representational techniques to the cognitive challenge that the hemispheric scale of the Pacific war presented to ordinary Americans. Their images offered Americans a vision of global war.[1]

In this chapter I examine the challenges to popular American geographical imagination and representation produced by the Pacific's emergence as a single geographic space within the geopolitics of competing imperial powers from the late nineteenth century to the mid-twentieth. I

185

do so by tracing the connections between Pacific mapping and the way Americans 'saw' this oceanic space during those decades, paying special attention to Pacific politics in the 1890s and 1940s, to the spatial impacts of technological changes in transportation and communications, and to 'popular' mapping in schoolbooks, newspapers and magazines. Owens' *LA Times* pictures are best understood as the outcome of a long history of attempts to envision and represent a global spatiality whose most demanding expression was the scale of the Pacific Ocean itself.

Charting Pacific space

The historiography of Pacific mapping and charting since Magellan's circumnavigation has conventionally been dominated by a narrative of discovery and culture contacts, of tracing of coastlines, accurately locating islands and atolls, and of reducing and finally dissolving into two island continents the ancient myth of a great southern land mass: *terra australis incognita* (see Fig. 3.4).[2] A minor tradition of study focuses on navigational history, from the (to Western eyes) remarkable 'stick' maps of Melanesian sailors and the haven-finding methods of the Manila Galleon, to the Pacific science, naval charting, art and ethnography that opened with the eighteenth century navigations by Cook, Bougainville and La Pérouse.[3] The Pacific as a bounded geographical – or, better, oceanic – space emerged relatively slowly, from the gradual extension of coastline charting and accurate fixing of island locations, a process that lasted until remarkably recently with the final delineation of Antarctica. Already split and pushed to the margins of the Western world image by the seventeenth-century Dutch innovation of the double-hemisphere map, Pacific space, even in the mid-nineteenth century, presented cartographic problems, whose (partial) resolution during subsequent decades shifted the narrative away from discovery and towards global geopolitics. The principal cartographic challenges concerned the hemispheric scale of the Pacific Ocean, the homogeneity of its oceanic space, and the time/space implications of defining East and West on an imperial globe. These matters remain in some respects vital today; between 1850 and 1945 they dominated Pacific spatiality.

Whaling was the principal early stimulus in shifting the view of the Pacific from a network of linear trading routes and expedition itineraries towards a territorialized space. In the 1830s and 1840s some 500 American whalers regularly joined those of the European nations, ranging across Pacific waters from Peru to Kamchatka and into the Bering Sea. Although largely spent by 1860, whaling had provided Westerners with a foundation of empirical knowledge, not only of differently productive whaling grounds, but of islands and atolls, and, more significantly, of oceanic winds and currents. Matthew Fontaine Maury was the first to utilize this information systematically in order to map the ocean's geography.[4] As first Director of the Depot of Charts and Instruments (later the US Hydrographic Survey) between 1839 and 1866, Maury had access to log books dating back to the previous century. In addition, he developed a standardized wind and current recording method for use by whaling, commercial and naval ships sailing in the Pacific. Applying statistical mapping techniques pioneered by Alexander von Humboldt, Maury produced the first wind and current chart of the Pacific in 1847. His maps allowed the ocean to be conceived and represented for the first time as an integral geographical space, and as a world region. Maury's maps would become familiar to American students through their reproduction in his geography texts which were widely used in American schools during the critical years of US expansion as an imperial Pacific power.[5]

Imperial science in the Pacific

Maury's work reflected an American involvement in the Pacific that was strengthening even as commercial whaling declined. Acquisition and settlement of Oregon and California in 1846–8 under the banner of Manifest Destiny gave added political significance to Charles Wilkes's and Cadwalader Ringgold's contemporary explorations in the Pacific. The scientific outcome of US naval expeditions included maritime atlases, and among their political consequences was American hegemony in Samoa. These explorations, enthusiastically promoted and scientifically supervised by Maury, might properly be termed 'imperial science'. They cannot be separated from Commodore Perry's Japanese expedition of 1852–4, which, among its various achievements, mapped potential coaling stations across the Pacific Ocean.[6] Navy Secretary John P. Kennedy's instructions to Perry made clear the expedition's geopolitical imperatives, placing it in the context of 'navigation of the ocean by steam, the acquisition and rapid settlement by this country of a vast territory on the Pacific, the discovery of gold in that region [and] the rapid communication established across the isthmus which separates the two oceans'.[7]

After a hiatus during the Civil War years, the final decades of the century witnessed US participation in an imperial struggle in the Pacific that secured the ocean in America's geopolitical imagination as a single global space. If the 'scramble' between Western powers and Japan was neither so dramatic nor so well documented as that for Africa in the same years, it was equally ruthless and geopolitically consequential. America's 'Open Door' policy, aimed at securing trading outlets in Asia, complemented the 1821 Monroe Doctrine (framed initially to limit Russian advance down the Pacific coast of

North America) and demanded a naval military presence in the ocean to secure coaling stations. Purchase of Alaska in 1868 removed Russia as a player in the Pacific imperial game, but Great Britain, France and above all Germany remained potential rivals. Again, science and mapping paralleled and complemented the political struggles that eventually yielded American control of Hawaii and Samoa. Deep-sea sounding surveys between 1872 and 1899, aimed at determining appropriate submarine cable routes, provided more detailed mappings of water depths and temperatures, ocean currents and winds, further enhancing the image of the Pacific as a unified space. The presence of US naval survey vessels projected the Republic as a Pacific power, feeding expansionist visions of American imperialists in the closing years of the century.

By the 1890s imperial science had produced a unitary geography of many physical features of the Pacific, which underpinned its significance as a geopolitical space. It had also given graphic expression to one of the key ideological themes in imperial discourse. In 1884 the International Meridian Conference held in Washington DC fixed the prime meridian at Greenwich. The implications of this decision for an International Date Line along the 180° meridian, which ran conveniently for most of its course through 'empty' ocean space, were acknowledged at the same conference, although this yielded no formal agreement on its status. The calendar anomalies inherited from the Renaissance treaties of Tordesillas and Zaragoza that had left the Philippines on American time and Alaska on Asian time were resolved between 1844 and 1892, so that the global division of East and West finally entered literate consciousness as a fixed line bisecting Pacific space, and henceforth clearly demarcated on globes, maps and atlases. The significance of this hemispheric division should not be underestimated at a time of violent anti-Asiatic nativism in American West Coast cities and discussions of a world-historic struggle between the 'Christian' West and Asiatic 'barbarism' among serious academics and political thinkers.

West and East: the geopolitical moment

In 1906 Frederick Jackson Turner, whose 1892 paper on the closing of the American frontier had itself closed with the tantalizing question of how American democracy was to be secured in its second century, returned to the significance of the western states in the context of outward expansion of the American frontier. Their settlement, he declared, constituted 'a call to the lodgement of American power on the ocean, the mastery of which is to determine the future relations of Asiatic and European civilization'.[8] With the seizure of the Philippines, Guam and Wake Islands in 1898, American colonial possessions now ringed the Pacific, with Pearl Harbor as the navel of an imperial ocean. Jackson's terminology echoed a broader fin de siècle discourse, to be found in imperialist writings in both Europe and America. Geopolitics, the argument that enduring geographic elements such as the distribution of lands, seas, climates, natural resources and populations determined the political history of humankind, arose in part from a widespread consciousness of 'closed space'.[9] Both scientific and political events in the Pacific at the turn of the twentieth century played a significant role in American versions of this discourse.

In 1885 Alfred Thayer Mahan (1840–1914), an American navy captain whose active service had coincided with the doldrum years of American investment in maritime power, was appointed to teach history at the newly established Naval War College at Newport, Rhode Island.[10] There he wrote the work that would propel him to international prominence as modern imperialism's key thinker on sea power. The Influence of Sea Power upon History (1600–1783) was published in 1890.[11] Its introduction and first chapter sought to establish enduring principles, from the idea of the sea as a 'great common', through other 'elements of sea power,' of which its author considered geographical position, physical conformation, territory, population and 'national character' as prime. Mahan, initially a reluctant imperialist, concluded in the text that colonies were an inevitable and necessary dimension of maritime power. In subsequent lectures and writings he became an ardent advocate of US naval power, especially in the Pacific, the Caribbean and the Isthmian zone whose canal finally linked them in the year of his death.

Mahan's emphasis on sea power represents one side of the fin de siècle geopolitical debate. The other side is to be found, perhaps ironically, in the writings of a British imperialist. Halford J. Mackinder's 'Geographical pivot of history', read to the Royal Geographical Society in 1904, in part a response to Mahan, argued for the world historical significance of Eurasian land power.[12] Mackinder's argument, captured in his famous map (Fig. 11.1), divided global space into a 'pivot area' inaccessible to naval penetration, an 'inner' or 'marginal crescent' and an 'outer or insular crescent' that he described as 'wholly oceanic'. His map significantly reduced the area of the Pacific relative both to the Eurasian land mass and to an empty ocean south of latitude 40°S, while relegating the Americas to the periphery of world history.

Despite their strategic differences, Mahan and Mackinder shared a common belief in a world-historical struggle of West and East, brought to a head in the closed space of imperial rivalry. The clash was framed in racial-religious terms as a struggle between 'white' European Christianity and 'yellow' Asiatic paganism. For Mackinder, 'European civilization is, in a very real sense, the outcome of a secular struggle against Asian invasion', framed in his dramatic metaphor as 'a blow . . . from the great Asiatic hammer striking freely through the vacant space'.[13] In the same year, a fellow

THE NATURAL SEATS OF POWER.

Pivot area—wholly continental. Outer crescent—wholly oceanic. Inner crescent—partly continental, partly oceanic.

Figure 11.1 Sir Halford Mackinder, Map of the heartland, ('The geographical pivot of history' [1904], in *Democratic Ideals and Reality: With Additional Papers*, New York: W. W. Norton, 1962)

Englishman, proclaiming the same world-historical struggle, pointed to the Pacific Ocean as the zone 'where seven empires meet [as] the battle ground on which will be fought the great racial struggle of the present'.[14] British 'Blue Water' strategists of the 1870s had ranked the Pacific as their nation's fourth most strategic trade route demanding naval protection. But it was an American who had first focused attention on the Pacific as the geopolitical space where East and West would finally collide. William Seward, who had effected the purchase of Alaska in 1867, had justified his 'folly' to the Senate by describing the Pacific as 'chief theater of events in the world's hereafter'. Americans, he had argued, were destined 'to roll its restless waves to the icy barriers of the north, and to encounter oriental civilization on the shores of the Pacific'.[15]

Mahan accepted this view of the Pacific as the space of an epochal encounter of civilizations. In his 1897 text, *The Interests of America in Sea Power*, he claimed that 'we stand at the opening of a period when the question is to be settled decisively, though the issue may long be delayed, whether Eastern or Western civilization is to dominate throughout the earth and to control its future. . . . What then', he asked, 'will be the actual conditions when these civilizations, of diverse origins and radically distinct, – because of the evolution of racial characteristics radically different – confront each other without the interposition of any neutral belt . . .?'[16] The Pacific, he suggested, had acted historically as such a belt. But America's 'open door' policy of equal commercial opportunity across its oceanic trade routes had ended that role with, in Mahan's view, disturbing consequences, as Chinese and Japanese labourers flooded into America's West Coast cities. In what reads today as a stunning inversion of logic, Mahan claims that 'a large preponderance of Asiatics in a given region is a real annexation, more effective than the political annexations against which the Monroe Doctrine was formulated'.[17] America's Pacific obligations were two-fold, to protect its strategic interests and to ensure the allegiance of Pacific peoples to 'Christian civilization'. These imperatives came together in Hawaii, 'an instance in point' of 'racial annexation'. In 'Hawaii and our future sea power' (1893) Mahan offers a graphic vision of the Pacific:

> To anyone viewing a map that shows the full extent of the Pacific Ocean, with its shores on either side, two striking circumstances will be apparent immediately. He will see at a glance that the Sandwich Islands stand by themselves, in a state of comparative isolation, amid a vast expanse of sea; and, again, that they form the center of a large circle whose radius is approximately – and very closely – the distance from Honolulu to San Francisco.[18]

This is a perfect example of one of geopolitical reasoning's characteristic features: using the world map to treat geographical space as pure geometry. Global race theory, a second characteristic feature of geopolitical thought, is also explicit in Mahan's writing:

> It is a question for the whole civilized world and not for the United States only, whether the Sandwich Islands, with their geographical and military importance, unrivalled by that of any other position in the North Pacific, shall in future be an outpost of European civilization, or of the comparative barbarism of China.[19]

'Race patriotism', to use Arthur Balfour's invidious phrase, placed the burden on America, 'from the point of view of the national interest as well as of . . . moral obligation, to assist the development of "alien subjects, still in race-childhood"', and thus to annex the islands.[20]

Rather than China, it would be Japan, whose people already dominated Hawaiian demography, and whose migration into the western states of the US so concerned fin de siècle American nativists, that would emerge as the principal competitor to American imperialism in the Pacific. After Japan's stunning naval victory over the Russian empire in 1905 and its colonial occupation of Korea in 1908, Japanese intellectuals consciously reworked Japanese identity along lines of ethnic nationalism borrowed from Western race theory. In so doing, they recast the Japanese geographical imagination, replacing the Chinese-originated term Yazhou as a designation for the geographical space within which the nation operated, with the term *toyo*.[21] This term, often translated as 'Orient', means literally 'eastern ocean', and its adoption in the early twentieth century implied Japanese distinction from and superiority over China's continental domination, as well as equivalence with European states: Japan projected itself as both modern and Oriental.[22] For many nationalists, this oceanic spatiality cast Japan in a parallel role to Great Britain in the Atlantic, justifying the geopolitics of Japanese imperial hegemony in the Pacific.

Popularizing an imperial Pacific map

Mahan's influence was at its peak at the time of the Spanish-American War of 1898, fought precisely in the zones he had signalled as America's maritime 'back yard': the Caribbean and Pacific. Victory was popular among ordinary Americans, whose cartographic and geographic literacy at this time was unusually high. At the turn of the twentieth century, geography was more widely studied than history in American schools; indeed geography partly subsumed historical study. Geography's essential pedagogical tool was the atlas. In fact, geographical texts were constructed as commentaries on mapped data, easily and cheaply reproduced.

> By the 1880s . . . American cartography had evolved from an elite craft to a mass industry producing maps as quickly and cheaply as possible. New technologies, organizational modes, and economic motives facilitated the spread of inexpensive atlases through American culture, a trend that made atlases a common feature in urban parlors and rural schoolrooms alike.[23]

We are familiar with the extent to which these atlases and school geography texts 'framed the world as a racial hierarchy by highlighting the unified relationship between race, climate and "progress", and in the process created an ethnographic world that functioned according to certain laws'.[24] Within these hierarchies, East Asian and especially Japanese people were ambiguously positioned:

> a constant source of fascination . . . At their best they were 'honest, ingenious, courteous, cleanly, frugal, animated by a strong love of knowledge, endowed with wonderful faculty of imitation, and possessing a sentiment of personal honor exceeding that of any nature'. But . . . they could also be 'fickle, inordinately vain, and, at least in the lower classes [not incidentally those present in American cities], exceedingly corrupt.'[25]

Such people could present genuine rivalry to 'Anglo-Saxon' dominance, while the peoples of the Philippines and Melanesia, dwellers of the torrid zone, remained among the ranks of 'mankind's children', thus suited to be the beneficiaries of the West's colonizing mission.

While these ethnographic claims were often illustrated by elaborate renderings of world climatic zones complete with flora, fauna and ethnographic stereotypes, it was maps of the world's physical geography that provided the foundation of geographical study. Indeed, the 1890s saw a renewed emphasis on global physical processes as the scientific basis for the distribution and characteristics of 'races'; these years mark the beginnings of American geography's three-decade flirtation with environmental determinism.[26] In this context it is worth examining the image of the Pacific offered to students in popular atlases and geography texts. Overwhelmingly these adopted Mercator's cylindrical projection, a map projection initially designed as a navigational device because straight lines drawn across it describe lines of constant compass bearing. Its advantages are gained, however, at the cost of radically distorting that area of lands and seas on the globe in favour of the higher latitudes. When combined with the tendency to centre atlas maps on the 0° meridian while cutting their latitudinal extent at 60–80°N and 50–65°S, the Pacific is significantly reduced in scale relative to other parts of the map.

An example of the global image offered to American students in the years of American Pacific imperialism is Arnold Guyot's *Physical Geography*, initially published in 1873 and revised in 1885.[27] Through his school texts, the Swiss-born Guyot remained among the most influential geographers in America for decades after his death in 1880. A firm disciple of Carl Ritter, his school text used a set of beautifully engraved and lithographed world maps to press 'the interrelationship between the land, the ocean, the atmosphere and humans, all of which interacted harmoniously in a grand design'.[28] The book's preface makes clear that these maps 'constitute in themselves a work as laborious as it is indispensable . . . selected strictly with a view to instruct the pupil, and not simply to adorn the pages of the volume'.[29] Apart from two small inserts showing land and water hemispheres, the maps uniformly adopt a Mercator projection with a latitudinal extent from 80°N to 65°S. Most maps are centred upon the Atlantic with a longitudinal cover of 430°, achieved by showing east Asia and Australia (110°E–180°E) at both edges. This allows the Pacific basin to be illustrated whole on most (but by no means all) of the distribution maps, although it remains marginal by comparison with the Atlantic Ocean (Fig. 11.2).

Guyot's principal competitor for the advanced physical geography market in school texts was *Maury's Physical Geography*, written by M. F. Maury himself, together with his family, after he resigned his federal post to support the Confederacy during the Civil War.[30] The text appeared in 1872 and was revised by his son in 1887.[31] Unsurprisingly perhaps, given his earlier

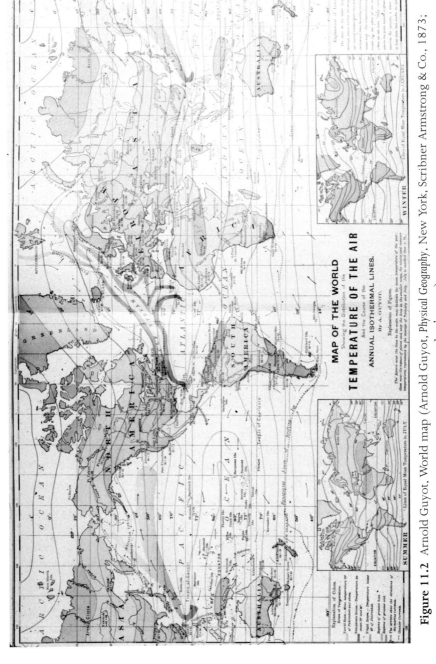

Figure 11.2 Arnold Guyot, World map (Arnold Guyot, *Physical Geography*, New York, Scribner Armstrong & Co., 1873; author's copy)

hydrographic work, Maury's approach to the world map is somewhat more adventurous than Guyot's. He rarely cuts the unity of the Pacific Ocean, and adopts the 120°W meridian as the centre line, making the map edges correspond to the supposed division of Europe and Asia along the Urals, extended south through Iran and the Indian Ocean. Bizarrely, one of the few maps that do divide the Pacific illustrates 'Currents of the sea'. But other than an opening double hemisphere 'on the globular projection', Maury, like Guyot, adopts Mercator for his world map, confining its longitudinal extent to 65° north and south. The true extent of the Pacific is thereby diminished even in a text that reproduces the work of the scientist most responsible for mapping the world ocean.

Both Guyot's and Maury's works pre-dated America's colonial expansion into the Pacific, and while each contained detailed illustrations of its physical geography, neither mapped the ocean in a manner that illustrated its true scale and significance on the globe. The dominance within American popular geography of the Mercator world map, reflected in both these works, was only seriously challenged as the struggle between America and Japan for imperial hegemony in the Pacific intensified.

Aerial visions of the Pacific between 1910 and 1945

Japan's 1905 victory over Russia set the stage for a potential geopolitical struggle with the United States in the Pacific that was not actualized until Japanese colonial expansion into Manchuria in the 1930s began to challenge America's Asian trading interests and security. For nearly four decades before 1941, American cartography had reflected a widely held view that 'the Pacific, or Oceania, [was] a distant group of scattered islands, a buffer *between* civilizations more than a society of its own'.[32] This was reflected in the near-universal use in interwar American schools and popular publications of the Mercator world map centred upon the Atlantic, visually reinforcing the perception of the Pacific as a barrier. Even John Goode's homosoline projection of 1923, developed to counter the distortions of a cylindrical projection, while accurately rendering the size of the Pacific Ocean, split it into two parts and pushed it to the periphery of global space.

Japan's challenge to America's Pacific hegemony became explicit in the 1930s. In 1931 Clyde Pangborn and Hugh Hendon, pilots of the first non-stop flight to cross the ocean, were arrested in Tokyo for landing illegally. Their flight and Charles Lindberg's great circle flight through Alaska and the Aleutians generated a renewed interest in oceanic spatiality. Pan American Airlines introduced a scheduled service between San Francisco and Manila in 1935 that took an island-hopping route via Hawaii and Midway, and its *China Clipper* was an object of widespread publicity and American pride.[33]

The popular writer and broadcaster Hendrik Willem van Loon's 1940 book, *The Story of the Pacific*, drew upon the popular interest in oceanic space.[34] Van Loon was already well known for his world history and geography books, *The Story of Mankind* and *Home of Mankind* respectively, and drew upon his reputation in the Pacific book to challenge American isolationism in the aftermath of Germany's invasion of his native country, the Netherlands. His text is based on personal experience of sailing through the Panama Canal and across the Pacific to the Dutch East Indies and Australia. In many respects it is a conventional narrative of European discovery and colonization of the Pacific, but it is characterized by van Loon's quirky humour, his humanism, his respect for non-European cultures, his distaste for imperialism – whether mercantile or missionary – and his sense of a coming conflict that would further waste the Edenic world he imagined had pre-existed Western entry into the Pacific:

> I wonder what . . . the result will be of that terrible chess game now being played among its islands and atolls by the powerful nations of Asia and Europe, all of them intent on acquiring new naval bases, new airplane bases, new army stations . . .[35]

Van Loon's answer came the following year at Pearl Harbor on 6 December 1941, jolting Americans' reassuring vision of invulnerability from Asia. The succeeding war produced a new Pacific spatiality, directly experienced by thousands of American troops engaged in combat across the ocean, and represented to their families and compatriots at home in a national effort of self-education in hemispheric geography. This effort generated original cartographic initiatives within American school geographies, to be sure, but it extended much further: into news journals, popular magazines and newspapers. The revolution in American spatial thinking about the Pacific, achieved between 1941 and 1945, was driven not only by the strategic realities of warfare, but by new technologies of conquering, mapping and picturing space and distance.

Airman's war, global vision

The Pearl Harbor attack was dramatically illustrated by Charles Owens in the *Los Angeles Times* of 8 December 1941. He captured graphically the American fleet's destruction, not from the perspective of American sailors aboard the stricken ships, but from that of a Japanese dive bomber swooping low over the harbour, placing his viewer at cockpit level, and observing the aerial attack as bombs and downed planes fell into the picture. The following week he visualized the attack from a lower angle, sketching the full fury of battle in a war-comic action-image of swooping aircraft, lines of tracer fire, explosions, sinking ships, flames and smoke. Owens' perspective is prescient, for the Pacific war was to be as much an air as a naval and land struggle; its opening and closing moments in Hawaii and Nagasaki would both be enacted from the air. And air power radically altered both geopolitical perspectives based on balances of land and sea power alone, and the spatiality of the Pacific Ocean.

This new dimension of warfare was recognized early on in the struggle, and gave rise to a sustained discussion among American intellectuals and strategists about the 'airman's war' and, more broadly, the 'airman's vision'.[36] Introducing a collection of geopolitical essays originally presented at an *Atlantic Monthly* symposium held to discuss the shape of the post-war world, the public intellectual Archibald MacLeish proclaimed the scope of a new spatiality offered by aerial vision:

> Never in all their history have men been able truly to conceive the world as one: a single sphere, a globe having the qualities of a globe, a round earth in which all the directions eventually meet, in which there is no center because every point, or none, is center – an equal earth which all men occupy as equals. The airman's earth, if free men make it, will be truly round: a globe in practice, not in theory.[37]

While the conflict in which the USA was then engaged was truly global, engaging American troops on every continent, no other single theatre of war approached the hemispheric scale of the Pacific. Here, the idea of an airman's war and an airman's vision was specifically realized. Existing global and Pacific cartography failed to capture this new spatiality. The American geographical imagination had to be reshaped if the war was to be understood and if, as MacLeish and his colleagues hoped, the peace was to be made on democratic terms that fulfilled America's promise to the world.

Those who pioneered new modes of cartography for capturing the airman's global vision for the Pacific conflict were not conventional mapmakers but journalists. Already in a 1936 colour insert in the *San Francisco Examiner* the journalist/artist Howard Burke had sought to give readers an idea of the strategic realities of the Japanese–American struggle in the Pacific. But it was the architect and artist Richard Edes Harrison who worked most closely with MacLeish and other members of the liberal establishment to capture the airman's vision for a wider American public. Harrison's maps were published in nationally circulating magazines, *Life* and *Fortune*, and soon collected into a single volume, the *Fortune Atlas for World Strategy*, entitled *Look at the World*[38] (Fig. 11.3). To capture the spatiality of the global conflict Harrison resurrected two well-known but rarely used projections: the polar azimuthal, which emphasized the closeness of the American continent to others in the 'land hemisphere', and the orthographic, which gives the viewer the impression of viewing the earth as a sphere rather than the

Figure 11.3 Richard Edes Harrison, Pictorial map of the Pacific ('*Look at the World*': *The Fortune Atlas for World Strategy*, Editors of Fortune, 1944; author's copy)

two-dimensional surface revealed in conical or cylindrical projections. Additionally, he developed a set of dramatic pictorial maps, artistic rather than mathematical renderings, to provide a perspective over the earth's surface as it might appear from an aircraft. In these, the surface is foreshortened and the shapes of land and sea masses distorted as if seen across the curving surface.

Harrison's orthographic map of the Pacific consciously seeks to capture both the scale and spatiality of the great ocean. Centred at longitude 170°E and latitude 35°N, the globe is tilted towards the northern part of the oceanic basin, thus emphasizing two strategic lines between Japan and the USA: the northern great circle route through the Kuriles and Aleutians to Alaska, and the vector of Japanese imperial expansion through Micronesia. A dotted line reveals the maximum limit of Japanese expansion, demonstrating Harrison's contention that 'the Japanese moved too late upon the vital northern great circle route. They were just too late at Midway; they were way too late at Kiska.'[39] He emphasizes the problems of grasping the scale of Pacific geography:

> Of all the great arenas of the world, the Pacific is the most difficult to visualize as a whole; it is the vastest single strategic unit in which men have ever struggled for mastery or survival. On a single hemispheric map it is impossible to show its east–west limits – for they stretch from the Isthmus of Panama, beyond the right edge of the map, to India,

198

shown on the far left. From one limit to the other is more than half way around the world. Furthermore, the north–south boundaries extend almost to the poles.[40]

The orthographic map is followed by three of Harrison's perspective maps, illustrating the airman's view over the Pacific to Japan: from Alaska, the Solomon Islands and southern India respectively. Dramatically coloured, they seek to capture enduring geographical relations between the ocean and its circumambient land masses.

A photographic reproduction of Harrison's orthographic perspective over the Pacific acted as a dramatic opening illustration to a special edition of *Life* magazine on 22 December 1941, accompanied by a caption describing the Pacific battlefield as covering half the planet. The *Life* article uses 20 pages of photographs and text to illustrate 'the oldest and greatest of the world's oceans, the vast and unresting water which rolls in from the west on the shores of America and breaks from the east at the coasts of Japan', now 'the vastest single battlefield over which man has ever fought'.[41] The opening image of global-scale emptiness is followed by photographs that capture the Pacific's unity and diversity: sunset over still water to illuminate the sun's nine-hour traverse of the ocean; great cities at the ocean's 'corners': Vancouver, Sydney, San Francisco, Yokohama, Vladivostok and Hong Kong (no South American cities are included); waves breaking in the Bering Straits or over islands and atolls in mid-ocean. The ethnographic commentary echoes in more condescending but equally offensive terms the verities of late nineteenth-century 'racial science':

> The winds and waters which move across the Pacific touch the homes of many kinds of peoples – the Japanese who cannot leave well enough alone, the Chinese whom a sense of long civilization has made more patient, the brown people of the Malay Peninsula and the islands, the white European settlers, the frizzle-haired primitives, the high-breasted girls and amiable men of the South Seas.[42]

The sharp division between East and West has, however, given way to an image of the Pacific as a global space with its own imperatives, a space that might be 'left well alone', but for Japan's aggression. The United States, chief Pacific imperialist, is the absent adjudicator of this ethnographic mosaic. The text of the *Life* piece deals in familiar tropes and stereotypes, addressing a readership raised on the geographical and historical ideas I have been examining. The appeal of the article, however, did not lie in the text, but rather in the photojournalism, and the pictorial Pacific was becoming very different from that of conventional geographies.

Charles Owens' Pacific maps

In the years before television, photojournalism vied with the movies as the principal graphic source of information in daily life. In daily newspapers, from the late 1920s on, photographs had begun to compete with artists' drawings and sketches to illustrate the news, but reproduction costs tended to restrict them to a single page, often page 3. It was page 3 of the *Los Angeles Times* that staff artist Charles Owens took over each Monday to illustrate the war's progress. Given the significance of the Pacific struggle to America, and specifically to Los Angeles, from which so many of the troops departed, illustrations of the oceanic conflict dominated his graphics. Owens was less of a trained cartographer than even Harrison, but early in his career he had developed a dramatically graphic way of presenting the geographical and spatial context of news events. Among his earliest news graphics were the San Francisco earthquake (1906) and the sinking of the Titanic (1910), both of which he illustrated by bird's eye views that combined distinctly cartographic elements (cardinal points, lines of longitude and latitude, distance measures, etc.) with pictorial renderings of the human drama. Inserts and blocks of explanatory text produced a complex, highly original, effective and self-sufficient pictorial news article, needing no accompanying narrative text.

The war images represent Owens' masterwork, drawing upon techniques he had honed over 30 years of news reporting in graphics. Like Harrison's, they seek to capture the Pacific as a single space of operations, but they are in some respects even more adventurous. While Harrison remained bound by such cartographic conventions as maintaining relative scale, accuracy of latitude and longitude and colouring relief, Owens' work was dominated by the goal of graphic communication, often at the expense of any pretence to scientific accuracy. His Pacific maps invariably show the curving horizon of a global perspective, regardless of their scale, be it the tiny Japanese atoll fortress of Truk or the entire Pacific basin (Fig. 11.4). In a Pacific map of 17 April 1944, the outline of the continental United States is overlain across the area of the Caroline Islands to demonstrate the vastness of the Pacific. 'Only by comparing the scope of these [wartime] undertakings with familiar distances does our war in the Pacific assume its real enormity', states the text.[43] Colour is used to indicate territorial possession and define combatant nations. Another oceanic map, from 21 May 1945, shows Japanese possessions in bloody red and their maritime zone in pink, while other territories (China, Australia, Russia, Burma and India) are rendered in more neutral tones. The ocean is always a vast space of pale blue. As on Harrison's map, the eastward limit of Japanese penetration is shown in a pecked red line, while the international date line separating East from West is given no prominence. Significantly, Owens' maps rarely show the full latitudinal

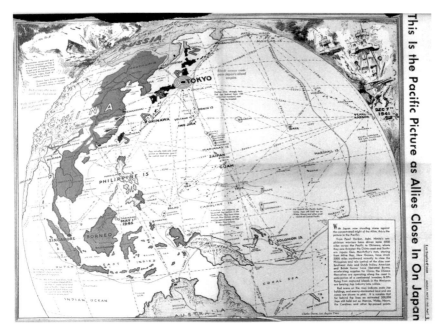

Figure 11.4 Charles H. Owens, The Pacific picture, *Los Angeles Times* war map, 1944
(UCLA Library)

extent of the Pacific; they focus on the battle zones west of Pearl Harbor. Yet Owens commonly exaggerates the curving global circumference to emphasize the scale of the Pacific's oceanic distances, a point constantly stressed in the text inserts.

The curving horizon line serves not only to focus the reader's attention on the global scale of the conflict, but also provides a space for Owens' battle scenes. On the Truk map of 13 December 1943, the scale of the atoll allows the artist to sketch in airfields, gun batteries, ships at harbour and even palm-fringed beaches. A tiny insert locates the atoll within the Pacific and arrows indicate distances. A dramatic headline tops the page: 'Truk, Japan's Ocean Fortress, Holds Key to Victory.' Between these words and the curving ocean horizon, Owens has sketched an intense naval and air battle, complete with sinking battle cruisers, dog fights in the darkening skies and explosions among the palm trees. In the bottom left of the map itself a fighter plane dives towards the atoll. The airman's view and the global vision are dramatically brought together in an image of Pacific wartime geography.

Conclusion

In the second half of the twentieth century, following America's victory over Japan in the Pacific, the ocean dropped rapidly off the radar screen of most Americans. The atomic tests at Bikini atoll, and Thor Heyerdahl's attempts to replicate early Polynesian voyages, both reinforced the image of the ocean's vastness, but at the expense of re-emphasizing its distance from everyday life. In American school atlases the Mercator projection regained its former hegemony, while the Cold War with the USSR focused attention on polar and European rather than Pacific geopolitics. With the independence of the Philippines (1946), the admission of Hawaii to the Union in 1958, and political independence for its various island groupings, Pacific space became carved up between national sovereignties rather than a geopolitical unit fought over by competing empires, and its mapping reflected this. It was satellite images and the dramatic colouring of El Niño ocean temperature effects that would reintroduce the Pacific to popular American consciousness in the 1990s – as an ecological space.

Mapping the Pacific geographically, that is, as a space in which physical forms, distributions and processes are integrated with human activities, presents unique technical and graphic as well as theoretical challenges. In twentieth-century Pacific mappings these challenges were closely connected and tied to the desire to educate a broad American citizenry about a strategic region. Fixing the date line at 180° had served to naturalize the Pacific as the world-historical fracture zone between West and East, figured as metonyms for civilization and barbarism, progress and decadence respectively. This spatiality superseded – while also incorporating – an earlier one that had translated the South Seas' distance from Europe into a historical and ethnographic regression from civilization.[44] In both cases, dividing the Pacific between the eastern and western edges of a cylindrical projection world map with a restricted latitudinal range at once marginalizes and diminishes the Pacific as a world region. Only in the middle years of the last century, during the decisive struggle for imperial dominance, did the Pacific become a unitary space cartographically and in the Western geographical vision.

12 Seeing the Equator

For whosoever once hath fastened
His foot thereon may never it secure
But wandreth evermore uncertain and unsure.[1]

A dimensionless line girding the midriff bulge of an oblate globe, the 40,070km Equator serves as a primary cosmographic trace marking geographical differentiation on the earthly sphere. The hemispheres that it simultaneously produces and separates are the largest units below the scale of the globe itself. Their symmetry has continuously influenced the ways that Earth has been imagined, understood and named since its sphericity was first recognized.[2] So powerful is *The Line* within the geographical imagination that the acts of 'seeing' it, or registering its crossing, do not, at first glance, appear as bizarre as on reflection they undoubtedly are. The Equator has many geographies: mathematical, physical, cultural, historical. This chapter focuses on its cultural geographies, and considers some of the ways in which the equatorial line has been envisioned and marked, along with the cognitive and cultural implications involved in these activities.

As we address the Equator's geography, its line expands into a region marked by natural features as dramatic and by places as distinct as any on Earth. The absence of any physical expression of the mathematical line itself coexists with the material presence of equatorial geography to generate a powerful imaginative mystique that finds expression in a complex equatorial symbology and recurrent tropes, and has generated a rich history of equatorial travel and writing.

Line and region: mathematical and elemental equators

Geography is not geometry; the Equator is both of these. As geometry, an equator is any diameter drawn on the sphere. Its location on the terrestrial globe is determined by the fact of the earth's rotation around its polar axis. The equatorial line is nothing more than the mid-point between the poles when measured across the spherical surface, 90° of latitude distant from each pole. It attains greater cosmographic – and environmental – significance from the geometrical relations between terrestrial and celestial motions. The night sky appears to the naked eye as a hemispheric dome slowly pivoting around us over the course of a year. Geocentrism, that intuitive belief in the cosmic centrality of the Earth, projects the earthly poles, Equator and hemispheres onto the celestial sphere. The 23.5° angle between the earth's axis of diurnal rotation and that of its annual rotation around the sun produces a line called the ecliptic. The ecliptic contains the zodiacal band in the heavens, defines the tropic lines of Cancer and Capricorn, and intersects the Equator at midday on the vernal and autumnal equinoxes. The apparent movement of the sun over the surface of the globe generates the geometry of five great circles that Greek cosmologists projected across both terrestrial and celestial spheres; of these the Equator has the greatest diameter.

The clarity of spherical geometry so apparent in the Equator stimulates a desire for similar symmetry between the hemispheres on either side. Hemispheres are by definition equal: north and south balance each other. So long as recorded knowledge of the globe remained the preserve of the northern hemisphere alone, its southern equivalent, as I have observed at various points in these essays, was imagined to resemble the northern in geography as well as in geometry. The Equator became effectively a mirror in which the northern part of the earth was reflected south. Thus the classical Greek and Roman geography that was later mapped into Arab and Christian images of the globe balanced the known and inhabited parts of the earth – the tri-continental world island or *oikoumene* – with a similar, if less corrugated, continent below the Equator. (see *Fig. 1.3*). Equatorial space itself was commonly shown as water: part of the ocean flow (*alveus Oceanis*) that circled the continents. *Terra Australis Incognita* – the unknown southern continent – retained a presence on world maps well into the eighteenth century (see *Fig. 9.1*), and endures today in the name of Australia, the only inhabited continent located wholly south of *The Line*.

Earth

The mathematical and astronomical relations that create the geometrical Equator do have physical and environmental consequences on the Earth's surface. The speed of the globe's rotation is greatest at the Equator: the consequent saving of fuel necessary for launching a missile into orbit is a principal reason for locating the French satellite launching station in Equatorial Guinea rather than in France proper. Sir Isaac Newton correctly predicted in the *Principia* that the gravitational effects of this variation in rotation necessitated a slight bulging of the terraqueous sphere such that the true form of the globe is an oblate spheroid: 'if the earth were not a little higher around the Equator than at the poles, the seas would subside at the poles, and by ascending at the region of the Equator would flood everything there.'[3] The French Scientific Academy's expeditions to Lapland and to Peru in 1735 and 1736, intended to resolve the dispute between Newton and Jacques Cassini over this matter by measuring the length of an arc of longitude, proved empirically the validity of Newton's claim. The Peru expedition was the first to make the Equator itself an object of scientific field research and has since been accorded a heroic place in the annals of Enlightenment science.

Fire

If the Equator thus finds physical expression in shaping the solid earth, it is even more apparent in the elements of fire, water and air. Together with the rotating globe, it is fire – in today's scientific parlance, solar radiation – that most characterizes the elemental geography of the Equator. Classical Mediterranean cosmography, accepted by both Arab and Latin scientists until well into the seventeenth century, divided the globe into five zones, symmetrically distributed around the Equator: two frigid zones, lying between the poles and the polar circles, two temperate zones, between the polar and tropic circles, and a single torrid zone located between the tropics. The Equator bisected the last of these. With the sun close to the zenith throughout the year, the tropic zone of fire was long deemed to be uninhabitable by humans, a belief with consequences that echo through to our own days in temperate attitudes to tropical regions and their peoples. The geography of these climatic bands, together with their ethnographic implications that those who did inhabit such spaces were somehow less than human, was only seriously challenged with the early nineteenth-century mapping of isotherms (lines of equal surface temperature) across the globe, and the moral arguments of the anti-slavery movement.

Isothermal mapping introduced a more complex cartography of world climatic regions than the traditional zones. The most influential classification of world climates has been that of the German geographer Wladimir Köppen (1846–1940), who in 1928 divided the torrid zone into northern and southern regions and inserted an 'Equatorial climate' between them, thus giving *The Line* a territorial expression in the earth's physical geography that remains widely accepted by scientists. The defining character of this belt is that mean daily temperatures do not vary by more than 3°C over the course of a year.[4] The geographical distribution of continents and oceans modifies the geometrical symmetry of the belt, so that it follows roughly the 'thermal equator', a line defined by joining the locations of highest mean annual temperature at each longitude.

Insolation and solar radiation levels in the zone of fire find expression in the forms and patterns of plant and animal life: the Equator is characterized above all by fecundity and thus by a speed and complexity of natural selection that has produced quite astonishing diversity of species by comparison with other regions of the earth. On the Equator in Borneo, 1000 tree species can be found in a single square kilometre of land, compared with a mere 700 in the 19 million temperate and frigid square kilometres of the North American continent, and it is estimated that the equatorial gene bank contains fully two-thirds of all living species.[5] It is thus ironic that European doctors long believed that women who crossed the Equator would become sterile.[6] There is a paradoxical relationship between equatorial fecundity and human disease that has always generated both desire and horror in peoples from more temperate regions of the earth. Long the locus of colonial sexual fantasies, famously played out by men such as Paul Gauguin and Richard Burton, and also the 'white man's grave', the Equator today is both the place where multiple sclerosis is unknown, and the originating location of such nightmare epidemics as AIDS and forms of necrotizing fasciitis (flesh-eating diseases). The popular geographical image of the Equator is still dominated by fire. Today it is the fires that annually reduce the supposedly fragile biodiversity along The Line to ashes, exposing that other paradox of equatorial geography: soil fertility is on average at its lowest here, and forest lands cleared for agriculture rapidly lose their productivity.

Water

As seafaring culture has disappeared from the daily experience of all but an ever-diminishing number of mariners, popular attention has become concentrated on the continents crossed by The Line.[7] Yet for more than four-fifths of its length the Equator passes through ocean, and it is in equatorial waters that the engine of global climate and thus of life on earth seems increasingly believed to be located. The movement of waters across the globe is of course an enormously complex and only partly understood function of rotation, thermal heating, convection gradients and winds. These generate currents that have only gradually and still imperfectly come to be known, mapped and understood by humans, from ancient Polynesian traders migrating between Pacific Islands, via nineteenth-century whalers' charts, to contemporary researchers examining the phenomena of El Niño and La Niña. The basic patterns of ocean currents are mirrored between the two hemispheres, while eastwards along the equatorial line itself moves a distinct flow of water whose temperature fluctuations have significant consequences for global climate. The equatorial current is immensely rich in marine plankton, its fecundity attracting some of the oceans' greatest creatures, creating a laboratory for the marine biologist and a source of awe and terror within the popular imagination. As Herman Melville, Jules Verne and others so well understood, equatorial seas are home to gigantic sea-creatures: the leviathan whale, the giant squid and the vast and ominous moray eel.

Air

The awe inspired by those strange life forms that occasionally arose from equatorial depths was, for mariners in the age of sail, exacerbated by the stillness of equatorial waters. The Equator is the region of the doldrums, where the global circulation that forms the so-called Hadley Cell warms and lifts air, producing a globe-girding corridor of low-pressure. And because the Coriolis force generated by the earth's circular motion has no measurable effect within five degrees either side of the line, no cyclonic winds are generated, nor do cyclones from more temperate latitudes move towards the Equator. The result is a zone of only light and variable winds, and very often no appreciable air movement at all for extended periods. Sailing ships famously could be becalmed near *The Line* for weeks, their crews broiled under a zenithal sun, their fresh water diminishing and their rations mouldering. Here, French slavers would poison their human cargo with quicklime before throwing them overboard, supposedly a more 'humane' response than the British and American practice of jettisoning live men and women to the mercies of waves and sea monsters, so vividly captured in J. M. W. Turner's *Slavers Throwing Overboard the Dead and Dying: Typhon Coming On* (1840).[8] So powerful was the hold of this maritime region and its conditions over the imagination of seagoing nations that 'the doldrums' remains a metaphor for psychological listlessness and depression today. With the coming of steam navigation the dread of equatorial seas was reduced, but the stillness of their waters remained a noted characteristic. Mark Twain, crossing *The Line* in 1897 as a passenger on a steam voyage from India to Mauritius, writes of the lake-like stillness of the ocean near the Equator:

> There is no weariness, no fatigue, no worry, no responsibility, no work, no depression of spirits. There is nothing like this serenity, this comfort, this peace, this contentment, to be found anywhere on land. If I had my way I would sail on forever and never go to live on the solid ground again.[9]

But still, equatorial seas can be whipped into sudden violence by the huge thunderstorms that are generated by the convective lift of moist air in equatorial heat. These are storms that do not move but rather exhaust themselves locally, but their internal winds can be as devastating as any other tempest. One of the ways that the Equator is visible on satellite images of

Earth is through the grey stain of storm stretching away from the western coast of Africa or South America.

Over the past half-century we have become familiar with images of the earth viewed from space. Indeed, their ubiquity has rendered obsolete conventional maps that marked the Equator, often with great emphasis, within the network of great circles, latitudes and meridians. Modern globes and world maps, influenced by satellite images, often show no mathematical geography, but allow the equatorial region to be identified as a region through changes in land surface colour (see Fig. 1.4). The Equator's visibility on maps and globes fluctuates, as it does on the physical earth. The desire to see the Equator can produce the sensation of actually doing so. Mark Twain wrote: 'Crossed the equator. In the distance it looked like a blue ribbon stretched across the ocean. Several passengers kodak'd it.'[10] And Thurston Clarke, introducing the record of his own, more recent, travels along The Line, notes that 'for centuries sailors have pasted a blue thread across spyglasses offered to shipmates for "viewing" the Equator. One nineteenth-century traveler reported cabin boys being "sent aloft to see the line." They came down describing a blue streak.'[11]

Continents and islands

The general geography of the Equator may be complemented by a specific geography of individual places along it.[12] Overwhelmingly, as I have pointed out, The Line crosses ocean, a higher proportion indeed of water to earth than that of the total global surface, for, despite the ancient desire for geographical symmetry, the distribution of lands on the globe is markedly skewed in favour of the northern hemisphere. Only two continents span the Equator: Africa and South America. South of Asia, the equatorial line crosses two of the earth's 'continental' islands – Borneo and Sumatra – cuts through the narrow promontories of Sulawesi and Maluku Utara, and crosses three closely neighbouring small islands in Irian Jaya. It is a remarkable fact that, beyond the few degrees of longitude that comprise this south-east Asian archipelago, only one island on the globe actually lies astride The Line. This is the northern part of Isla Isabella in the Galapagos group. Even the Line Islands of Kiribati, named for their proximity to the Equator, escape this distinction; tiny atolls in the Gilbert Island group lie either side of The Line, but none sits astride it.

Continent

Equally remarkable is that in both Africa and South America the Equator is marked by some of the highest peaks on the two continents as well as the low deltas of their greatest rivers, and in Africa by its largest lake. While Mount Kenya in Africa (5199m) and Mount Cayambe (5790m) in Ecuador are not the highest points on their respective continents, they are close competitors for this honour, and the upper slopes of each mountain are crossed by the equatorial line. The Equator traverses the Amazon delta's southern arm, and while the great Congo river enters the ocean at about 6° south latitude, the river's course crosses the Equator twice, and parallels it within 2° for over 1000km.

The altitude of the equatorial mountains produces the strange phenomena of snowfields, ice and glaciers at the mid-point of the torrid zone, a powerful inversion of climatic expectations. From ancient times it was known that the Nile's course crossed the Equator and that the constancy of its flow through vast regions of desert, as well as its mysterious flooding at the end of the dry northern hemisphere summer, were somehow related to this fact and to half-legendary mountains and lakes in the torrid zone. The famed 'Mountains of the Moon,' source of Egypt's great river, were generally placed well south of the Equator, but on Ptolemaic maps the equinoctial line itself was marked by Mount Amara and the nearby Lake Barcena within the Ethiopian kingdom of Prester John, whence flowed a Nile tributary. To Enlightenment scientists, equatorial mountains were equally fascinating. The altitudinal occurrence at Mount Chimborazo in Ecuador of the Earth's entire climatic and vegetational range was famously mapped by Alexander von Humboldt (Fig.12.1),[13] while the drama of Ecuador's sublime equatorial volcanoes inspired the vast romantic canvases of Humboldt's admirer, the American artist Frederic Edwin Church.[14] Mount Kenya, whose first reported sighting in 1849 by a European was dismissed on the grounds that snow could not exist at the African Equator, became, 50 years later, the site of a different kind of scientific practice. The first ascent was made by Sir Halford Mackinder in 1899, with the intention, it is often claimed, of cementing the Oxford Reader of Geography's credentials in the eyes of a sceptical Royal Geographical Society, then dominated by the cult of manly exploration and imperial conquest at the far-flung ends of earth.[15]

The equatorial lowlands of Africa, South America and the great islands of the South-western Pacific have played an equally significant role in the evolution of geographical science among temperate peoples. The classical idea of an uninhabitable torrid zone died very hard. The Renaissance iconography of Africa, created when the Portuguese were revealing – and ruthlessly exploiting – the human population that did live on the Equator, is telling. A naked black woman carries an umbrella to shield her from the pernicious rays of the zenithal sun that was long thought to produce haemorrhage of the brain after even minimal exposure. America's (that is South America's) iconography of the same years illustrates the other enduring

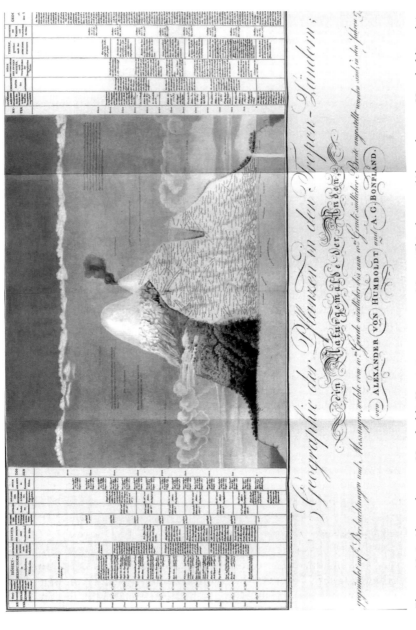

Figure 12.1 Alexander von Humboldt, Vegetation zones on Mount Chimborazo (Alexander von Humboldt and A. G. Bonpland, Geographie der Pflanzen in den Tropenländern, ein Naturgemälde der Anden, gegründet auf Beobachtungen und Messungen, welche vom 10. Grade nördlicher bis zum 10. Grade südlicher Breite angestellt worden sind, in den Jahren 1799 bis 1803, Staatsbibliothek zu Berlin, Kartenabteilung).

myth of these regions, applied equally to equatorial Africa, Borneo and New Guinea: that the Equator is the home of *anthropophagi*, or cannibals.[16]

The promiscuous density of plant and animal life in the humid, deltaic lowlands and rainforests of the Congo, the Amazon and the continental Pacific islands generates geographic and scientific descriptions that never seem fully to escape the fusion of powerful fears and desires with disinterested observation and wholly rational reflection. The modern study of 'tropical medicine' owes its origins (and many of its theories) to long-standing environmental beliefs.[17] Both Darwin and Wallace refined their scientific visions of life's self-generating and self-differentiating force – evolution – in equatorial regions, and their writings are deeply coloured by their awed experience of equatorial places. Such Victorian visions of the power of life at the Equator remained shadowed by equally compelling inscriptions of violence and death there. Once read, it is impossible to forget Joseph Conrad's descriptions of the horrors of Leopold's Congolese imperialism, or accounts of the terrible violence done to people and nature by rubber barons in early twentieth-century Amazonia.[18] The hold of the equatorial rainforest on our own geographical imaginations is no less powerful, as science allots to the climate, vegetation and biogeography of such places ever-greater significance for the environmental future of the whole globe.

A partial exception to the stark geography of mountain peaks and swampy rainforests is perhaps the one region of the continental Equator where temperate outsiders found an environment they believed they could understand, control and settle. On the plateaus of Kenya, above the great African Rift Valley, Edwardian British settlers produced a unique colonial landscape on territory leased from the Kikuyu people, whose land holding and use patterns they signally failed to understand or respect. It was a landscape of large-scale commercial farming, recreational big-game hunting, muscular mountaineering, casual racism and self-consciously exclusive social life, captured in the now-ironic phrase, 'White Highlands'. It was a short-lived paradise for those who enjoyed it. Already in the 1920s the colonial government in London acknowledged the primary claims to Kenya of African peoples over European settlers, and by the 1950s these were being asserted ever more vocally by educated Kikuyus, the more radical of whom formed the Mau Mau secret society to fight for independence. The savagery with which the settler society responded to the relatively few violent attacks by this group on their persons and privileges fits predictably into the storybook of equatorial horrors. Fifty years after Kenyan independence was achieved, a simulacrum of the colonial White Highlands continues to be played out in the nostalgic tourism of equatorial social clubs, hunting lodges and veranda hotels.

Island

A similar conflation of exuberant life and the dark shadow of violent death plays out in visions of the equatorial island. Here, too, the classic moment is the nineteenth century, when the Equator represented the region of greatest European colonial activity. In the wake of Enlightenment navigators – Cook, Bligh, Bougainville and La Pérouse among them – European and American whalers found the Equator a prime location for their prey, as Herman Melville's Captain Ahab recognized as he crossed and re-crossed *The Line* in pursuit of the white whale. Utopian dreams, violence and death in the equatorial South Seas pre-date Europeans, as is now revealed in the unravelling history of status rivalry and over-consumption that erected Easter Island's carved monoliths while deforesting its land and finally destroying its people.[19] The arrival of temperate mariners, misfits and missionaries simply introduced new complications and a more detailed historical record.

The utopian aspect of oceanic cultural geography at the Equator draws upon the deep reservoir of European island imaginaries.[20] Thomas More's original *Utopia* of 1516 (located in the South Seas, if not exactly on the Equator) initiated a tradition of placing imaginary ideal communities in the tropics. Actual islands were inevitably commissioned to fulfil insular fantasies. Jodocus Hondius' 1606 map of the Guinea coast of Africa, *Guineae Nova Descriptio*, contains an inset of São Tomé, the island in the Gulf of Guinea whose offshore southern rocks are actually touched by the Equator (Fig. 12.2). Hondius represents the island as a perfect circle (it is actually an ellipse whose axis runs north–south), precisely bisected by *The Line*, producing a geometrical figure of utopia at the middle of the earth. São Tomé's recorded history incorporates both sunshine and *noir* sides of insular geography: it was a principal refuge for Jews expelled from Spain in 1492, but also the birthplace of the slave-based plantation system of cultivation (in this case sugar) which characterized European colonial exploitation at the Equator.

The myth of the South Sea island, located on or near the Equator, as a self-contained environmental paradise where a perfect (generally leisured) life may be enjoyed, endures through the Renaissance and Enlightenment into our own days. It has been elaborated in novels, films, tourist advertising and 'reality' television. Commonly, the myth is powerfully erotic, founded in equal measure on ideas of heat and equatorial fecundity and on the exotic 'otherness' attributed to tropical peoples:

> At the equator, the concepts of human, animal, vegetable and mineral lose their meaning and fuse together. The sun passing overhead, bringing light and shade, ingrains the realization that nothing is defined . . . a celebration without shame or innocence, of life and its liquid surge between the thighs.[21]

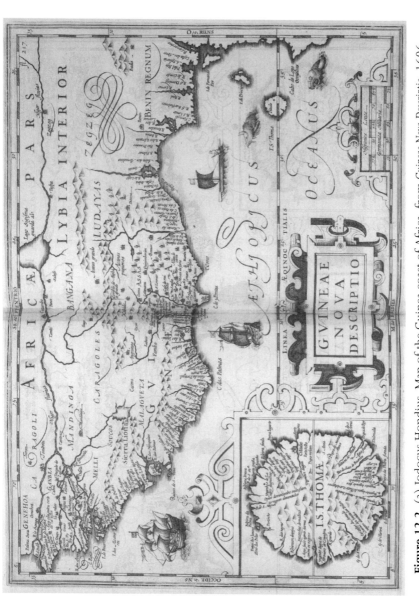

Figure 12.2 (a) Jodocus Hondius, Map of the Guinea coast of Africa, from *Guineae Nova Descriptio*, 1606 (British Library).

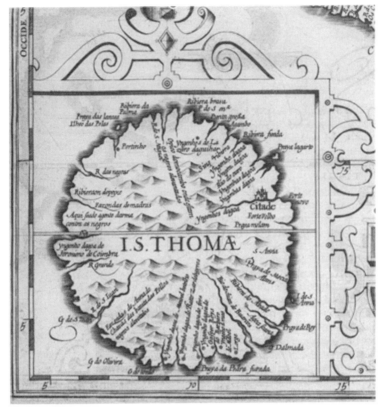

Figure 12.2 (b) Detail: the São Tomé insert in the map.

Various attempts have been made to realize paradise on the equatorial island. For four months in 1889, the Scottish writer Robert Louis Stevenson lived on Abemama and Kuria, the island atolls in the Gilbert Islands located nearest to the Equator. In the South Seas records both the invalid writer's personal attempts to live out the fantasy of a Pacific island paradise, and the savage debauchery of its native ruler, Tem Binoka. In 1938 the last formal British colonial settlement was established in the Phoenix Islands, a few degrees below the Equator, a failed attempt which was abandoned in 1963.

Violence and savagery continued to characterize the Gilberts throughout the twentieth century, fulfilling the dark counter-narrative that seems always to accompany the utopian. The 1941 Japanese occupation of Tarawa Island was followed by an appallingly misdirected American invasion during low tide at Betio to recover the atoll. Thousands of US Marines were slaughtered crossing razor-sharp corals to dislodge a tenacious enemy which then chose its own slaughter over surrender. In the subsequent decades Christmas Island, at 250km distant the closest to the Equator of Kiribati's Line Islands, was twice the location for atmospheric nuclear testing. In 1958 the island's British rulers used the mantle of International Geophysical Year to explode its *Grapple* series of air-dropped hydrogen bombs some 50km south of the island. Four years later, as the Test Ban Treaty talks between the USA and USSR faltered, the American Navy was granted permission to explode even more powerful devices above the island. The consequences for the health of its inhabitants, conscripted servicemen and the natural environment of Christmas Island have never been fully calculated or acknowledged.[22]

Today, the hundreds of tiny atolls that sprinkle the equatorial Pacific face similarly apocalyptic challenges. Many stand less than two metres above the tides, rendering their very existence vulnerable to the rise in sea levels commonly predicted by models of climatic warming. They are devastated environmentally by nuclear testing and phosphate extraction, their human populations battered by imported diseases, radiation and diet-related illness; the shadows of equatorial geography lie much longer across them than the luminescence of equatorial light.

The dark side of equatorial island existence dominates Herman Melville's geography of the Galapagos Islands, which bestride *The Line* in the eastern Pacific. *The Encantadas, or Enchanted Isles* (1854) comprises ten sketches based on the author's 1841 visit to this then isolated spot, six years after Charles Darwin studied the islands' uniquely evolved fauna. Melville's title evokes the myth of an island paradise, with echoes from Homer, Camões, Milton and Defoe, only to demolish it. His islands, 'five-and-twenty heaps of cinders dumped here and there in an outside city lot', suffer the

> special curse . . . that to them change never comes; neither the change of season nor of sorrows. Cut by the Equator, they know not autumn and they know not spring; while already reduced to the lees of fire, ruin itself can work little more upon them. The showers refresh the deserts, but in these isles, rain never falls. Like split Syrian gourds, left withering in the sun, they are cracked by an everlasting drought beneath a torrid sky.[23]

The islands are uninhabitable – 'man and wolf alike disown them' – and their principal occupants are reptilian: 'the chief sound of life here is a hiss.'[24] Their best-known inhabitants, the giant tortoises, are to Melville further grotesque evidence of evil enchantment, 'crawled forth from the foundations of the earth'.[25] Dark skies, violent seas and unpredictable currents add to the horror of Melville's Galapagos: 'Apples of Sodom, after touching, seem these isles.'[26] From the Roca Redonda, an isolated point that rises 80m from the ocean some kilometres north-west of the main Galapagos, he sketches a broad geography of the Pacific and of the islands themselves. The nearest and largest of them is Albermarle (Isla Isabela), 'a crater-shaped headland . . . cut by the Equator as exactly as a knife cuts through the centre of a pumpkin pie'. Its spatial symmetry is further enhanced: 'In the black jaws of Albermarle' lies Narborough . . . the loftiest land of the cluster . . . like a wolf's red tongue in his open mouth.'[27] At the heart of the equatorial island in this account is a darkness as black as Conrad's continental interior.

Travels at the Equator

To the many thousands of today's eco-tourists who visit the Galapagos each year, or to the countless others who know the equatorial islands through film and television documentaries, Melville's infernal account would no doubt come as a shock. The equatorial narrative has shifted. The Galapagos are now regarded as a unique, wondrous and vulnerable wilderness where the human footprint threatens a paradise constituted by nature's own intricate and creative agency. The newer narrative dominates the perspective of those who in increasing numbers seek to encounter the Equator for themselves, wherever they choose to do so.

'Did you ever lay eye on the real genuine Equator? Have you ever, in the largest sense, toed the Line?'[28] Melville's question is a compelling one. It reflects the power of the dimensionless Equator to become a destination as geographically present as any mountain, river, city or monument. The largest single body of contemporary literature devoted to the Equator consists of travel accounts. Dominantly, these are records of crossing the Equator; far fewer have 'toed the Line' in the sense of following along its path. Crossing the Equator is today a commonplace, although the act still seems to merit special note. A web search under 'Equator' will yield scores of blogs posted by travellers illustrating the moment of their crossing. At sea, tourist cruises seem to have revived the tradition (largely abandoned by the 1890s, according to Mark Twain) of first-time crossers being dunked in sea water by Neptune while drinking copious amounts of alcohol. On land, the iconic act seems to be placing a foot in each hemisphere. Even today, there exist relatively few points on the globe where a marked road or railway line crosses the Equator. Increasingly, those places are denoted by monuments and signs, attempting with varying degrees of success to exploit the economic potential of tourists wishing to mark their passage.

By far the largest and most visited of equatorial monuments is the Mitad del Mundo outside Quito in Ecuador, a state whose name, national identity and iconography are constructed from its location on the Equator.[29] The Quito monument thus signifies much more than a tourist attraction. The present 30m high pink stone obelisk topped with a concrete globe dates from 1986 and replaces a smaller 1936 version. Each was constructed to commemorate an anniversary of the 1736 French scientific expedition, and a complex of statues, museums, exhibitions and performances, as well as gift and souvenir shops, has now accreted to the monument. Other markers of the Equator are less impressive, in many locations no more than a roadside sign, although the recurrent iconography of obelisk, line and sphere denotes the fusion of geometry and geography. Thurston Clarke's 1980s account of his attempts to follow the line of the Equator around the globe contains descriptions of many of these monuments: concrete obelisks and spheres in Mbandka (formerly the explorer Stanley's Equatorville, in the Democratic Republic of Congo), Kotoalam and Bonjol in Sumatra, and the now destroyed obelisk near Kismaayo, Somalia, which was more a monument to Benito Mussolini's megalomania than to the Equator. Perhaps the most endearing of the equatorial crossings he records is the perfectly preserved colonial railway station at Equator, Kenya: 'A line of whitewashed bricks ran across the dirt platform, stopping at a globe mounted on a pedestal bearing a sign "EQUATOR. ALTITUDE 8,817 FEET".'[30]

Clarke's equatorial travels are unusual in consciously seeking to follow rather than cross the Equator, although he found this impossible in practice and was forced to make a series of visits along short stretches of *The Line*. He also avoids the common trope of muscular and masculine adventure that dominates equatorial travel literature, seeking instead to present a more peopled Equator, whose inhabitants he treats with respect and sympathy (although he does not escape the tropes of disease, violence and decay that temperate observers insist on reading into its lands and peoples). Climbing equatorial mountains, flying over its lakes and swamps, shooting its signature fauna (with gun or camera), together with the banal acts of transporting aquavit from the poles to the Equator and back, standing with one foot in each hemisphere, dunking one's head in sea water, or watching water drain down plugholes, have always been the more common ways that those who do not live on *The Line* have sought to register their encounters.

Conclusion

For a geographer, the Equator is a foundational location. It signifies the primary geographic project of bringing the variety and plenitude of the earth's surface into conceptual and classificatory order. Like the Earth's poles, the Equator is an absent presence that, once conceptualized on the theoretical globe, assumes a powerful significance as a physical location, to be located, mapped, experienced. Their physical non-existence enhances both the imaginative appeal of such geographical locations and the desire to 'see' them, or at least envision what exists there. A principal activity of geographical societies in the nineteenth century was to organize expeditions to the poles and Equator, and this tradition has not entirely ceased. Like the poles, the Equator has an eschatological presence: it is a place of origins and endings, an 'end of earth'. There we locate both myths of origin and fears of destruction: environmental, bacteriological and social.

The Equator thus encapsulates in powerful ways both senses of geographical vision used throughout this book. In the literal sense of seeing and representing actual places and landscapes across the surface of the earth and accounting for their difference and diversity, the Equator reminds us of geography's continuous dialogue between physical observation and graphic representation, between the field and the laboratory, studio or classroom, and still markedly, through the medium of the map, its task to 'make visible'. In the figurative sense of vision as a projective act of the imagination associated with desire or dread, the Equator has continuously acted as a reservoir of dreams for those living in cooler and more temperate parts of the globe, of visions that have not infrequently been the basis for action, with material consequences for actual places, landscapes and environments along the Equator. Thus, like every part of the earth, the Equator may be surveyed and mapped, photographed and remotely sensed, but its geography remains an open question, continuously available for further vision and inscription.

Notes

Introduction: Landscape, map and vision

1 Carl Ortwin Sauer, *The Morphology of Landscape* (Berkeley, CA: University of California Press, [1925] 1938). This famous essay has been reprinted and commented on in John Agnew, David Livingstone and Alisdair Rogers, *Human Geography: An Essential Anthology* (Oxford: Blackwell, 1996). See also John Leighly (ed.), *Land and Life: A Selection from the Writings of Carl Ortwin Sauer* (Berkeley, CA: University of California Press, 1963), especially Chapter 19, 'The education of a geographer', pp.389–404.

2 For example, Volume 6 of the series of conversations on key art historical concepts edited by James Elkins and published by Routledge is devoted to landscape theory with contributions from a wide, interdisciplinary group (*Landscape Theory*, London: Routledge, 2007). Other volumes deal with such subjects as aesthetics, global art history and criticism. Landscape architecture has been transformed in the past decade through conceptual debates about its object of concern, while both anthropology and archaeology have turned to landscape as a useful concept for grasping relations between culture, space and environment in contextual and synthetic ways.

3 'Maps and mapping images were frequently portrayed as prominent features of works of art, and often as wall decorations in the scenes captured on canvas. Indeed the appearance of such maps in such artistic works has suggested to some that they performed an authenticating role for the artists themselves' (David Livingstone, *The Geographical Tradition*, Oxford: Blackwell, 1992, p.99). Livingstone uses Jan Vermeer's famous painting of *The Geographer* as the cover image for his book. The painting, and its twin, *The Astronomer*, are discussed in Stephen Bann, *The Inventions of History: Essays on the Representations of the Past* (Manchester: Manchester University Press, 1990). See also Edward Casey, *Representing Place: Landscape Painting and Maps* (Minneapolis: University of Minnesota Press, 2002), pp.31–3 passim.

4 Casey, *Representing Place*.

5 Kenneth Olwig, *Landscape, Nature and the Body Politic: From Britain's Renaissance to America's New World* (Madison, WI: University of Wisconsin Press, 2002); Denis Cosgrove, *Social Formation and Symbolic Landscape* (London: Croom Helm, 1984).

6 In both Europe and the United States cultural geographers until the 1980s shared an interest in rural societies and landscapes, in folkways and 'tradition',

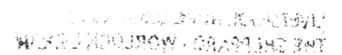

with an implicit distrust of modernization and urban industrial life as solvents of such supposedly stable ecological relations between environment and society.

7 On *Landschaft* and nationalism in Nazi Germany, see David Blackbourn, *The Conquest of Nature: Water, Landscape, and the Making of Modern Germany* (New York: W. W. Norton, 2006). For countries in eastern Europe seeking to redefine identities in the post-communist era landscape often plays a very significant role in national discourse. This is very clear in the case of Estonia. See for example 'My Estonia', an intensely patriotic video presentation of Estonian landscape produced by Estonian cultural geographers Hannes Palang and Helen Soovali (*Minu Eestimaa*, Sonatiin OÜ, 2005); Hannes Palang et al. (eds), *European Rural Landscapes: Persistence and Change in a Globalising Environment* (Boston: Kluwer Academic Publishers, 2004).

8 Martin Warnke, *Political Landscape: The Art History of Nature* (London: Reaktion Books, 1995); Simon Schama, *Landscape and Memory* (New York: Alfred A. Knopf, 1995). This is despite attempts to argue that landscape, like place, is 'process'. The claim is significant, but confused. Certainly landscapes are both active agents and material outcomes of social processes, but to restrict the meaning of landscapes (or places) to process is to dangerously underplay their materiality.

9 Denis Cosgrove, 'Epistemology, geography, and cartography: Matthew Edney on Brian Harley's cartographic theories', paper presented at the Association of American Geographers Annual Meeting, Chicago: March, 2006, published in *Annals of the Association of American Geographers* 97/1 (2007): 202–9; Arthur Robinson, *The Look of Maps: An Examination of Cartographic Design* (Madison, WI: University of Wisconsin Press, 1952); Arthur Robinson and Barabara Bartz Petchenik, *The Nature of Maps: Essays toward Understanding Maps and Mapping* (Chicago: Chicago University Press, 1976).

10 John Brian Harley, 'Maps, knowledge and power', in Denis Cosgrove and Stephen Daniels (eds), *The Iconography of Landscape* (Cambridge: Cambridge University Press, 1988), pp.277–312; John Brian Harley, 'Deconstructing the map', *Cartographica* 26 (1989): 1–20; John Pickles, 'Texts, hermeneutics and propaganda maps', in Trevor Barnes and James Duncan (eds), *Writing Worlds: Discourse, Text, and Metaphor in the Representation of Landscape* (London and New York: Routledge), pp.193–230.

11 Stephen Daniels and Denis Cosgrove, 'Iconography and landscape', in Denis Cosgrove and Stephen Daniels (eds), *The Iconography of Landscape: Essays on the Symbolic Representation, Design, and Use of Past Environments* (Cambridge: Cambridge University Press, 1988), pp.1–11. For critiques within landscape geography see Gillian Rose, *Feminism and Geography: The Limits of Geographical Knowledge* (Cambridge: Polity Press, 1993); Hannah Macpherson, 'Landscape's ocular-centrism – and beyond?' in Baerbel Tress, Gary Fry and Paulas Opdam (eds), *From Landscape Research to Landscape Planning: Aspects of Integration, Education and Application* (Boston: Springer/ Kluwer Academic, 2005), pp.95–104. Within art history, see W. J. T. Mitchell, *Picture Theory: Essays on Visual and Verbal Representation* (Chicago and London: University of Chicago Press, 1994), especially his idea of the 'metapicture' and his analysis of Peter Blake's art. A concern for the loss of the picture surface resulting from critical theory is being expressed among young art historians, for example Michael Gaudio, 'At the mouth of the cave: Thomas Cole and the

origins of American landscape painting', paper presented at 'The Art Seminar', 17 June 2006, at the Burren College of Art, Ballyvaughan, Ireland.

12 Vidal de la Blache's *Tableau de la géographie de la France* (1903) and the regional monographs that followed made extensive use of the IGN 1:50,000 and 1:100,000 topographic maps as sources for studying landscape and settlement morphology. The regional survey movement pioneered by Patrick Geddes and widely influential in the United States and Latin America in the early twentieth century also regarded the map as fundamental to geographical understanding. This is discussed more fully in Chapter 9.

13 The phrase 'non-representational theory' was used for the first time in the mid-1990s by Nigel Thrift as a theory of mobile practices, geared towards action, rather than contemplation: Nigel Thrift, *Spatial Formations* (London: Sage, 1996). Since then, new meanings have been attached and the phrase has more recently functioned as a self-critique in cultural geography. See Hayden Lorimer, 'Cultural geography: the busyness of being 'more-than-representational', *Progress in Human Geography* 29 (2005): 83; John David Dewsbury, Paul Harrison, Mitch Rose and John Wylie, 'Introduction: enacting geographies', *Geoforum* 33 (2002): 437; Ron Johnston et al. (eds), *The Dictionary of Human Geography* (Oxford: Blackwell, 2000), 556.

14 Martin Jay, 'Scopic regimes of modernity', in Martin Jay, *Force Fields: Between Intellectual History and Cultural Critique* (New York: Routledge, 1993); Macpherson, 'Landscape's ocular-centrism'.

15 Gillian Rose, 'Making space for the female subject of feminism: the spatial subversion of Holzer, Kruger and Sherman', in Nigel Thrift and Steve Pile, *Mapping the Subject* (London and New York: Routledge, 1995), 332–54; Derek Gregory, *Geographical Imaginations* (Oxford: Blackwell, 1994).

16 Reginald Golledge, 'Geography and the disabled: a survey with special reference to vision impaired and blind populations', *Transactions of the Institute of British Geographers* 18 (1992): 63–85; R. Butler and S. Bowlby, 'Bodies and spaces: an exploration of disabled people's experiences of public space', *Environment and Planning D: Society and Space* 15 (1997): 411–33; Hannah Hill, 'Bound to the environment: towards a phenomenology of sightlessness', in Robert Mugaerauer (ed.), *Dwelling, Place and Environment* (Lancaster: Nijhoff, 1985), 99–111.

17 See, for example, Halford Mackinder's famous ascent of Kilimanjaro. Gerry Kearns, 'The imperial subject: geography and travel in the work of Mary Kingsley and Halford Mackinder', *Transactions of the Institute of British Geographers* 22 (1997): 450–72; Geroid O' Tuathail, *Critical Geopolitics* (Minneapolis: University of Minnesota Press, 1996), 25–6.

18 David Livingstone, *Putting Science in Its Place* (Chicago: University of Chicago Press, 2004).

19 William A. Koelsch, 'Squinting Back at Strabo', *Geographical Review* 94 (2004): 502–18.

20 Daniela Dueck, Hugh Lindsay and Sarah Pothecary, *Strabo's Cultural Geography: The Making of a Kolossourgia* (Cambridge: Cambridge University Press, 2005).

21 Sarah Pothecary, 'Kolossourgia: a colossal statue of a work', in Dueck et al., *Strabo's Cultural Geography*, pp. 5–26.

22 Edward Said, *Orientalism* (London: Routledge and Kegan Paul, 1978); Gregory, *Geographical Imaginations*.

23 John K. Wright, 'Terrae Incognitae: the place of imagination in geography', *Annals of the Association of American Geographers* 37 (1947): 1–15; David Lowenthal and Martin Bowden (eds), *Geographies of the Mind: Essays in Historical Geosophy in Honor of John Kirtland Wright* (New York: Oxford University Press, 1976).

24 The key paper bridging Wright's ideas with those of post-positivist geographers was undoubtedly David Lowenthal's 'Geography, experience and imagination: towards a geographical epistemology', *Annals of the Association of American Geographers* 51 (1961): 241–60.

25 Paul Adams et al. (eds), *Textures of Place: Exploring Humanist Geographies* (Minneapolis: University of Minnesota Press, 2001); Steve Pile, *Real Cities: Modernity, Space and the Phantasmagorias of City Life* (London: Sage, 2005).

26 Elaborating this theme has been the singular contribution of Yi-Fu Tuan in a suite of books and essays, notably *The Good Life* (Madison, WI: University of Wisconsin Press, 1986). See also Robert David Sack, *A Geographical Guide to the Real and the Good* (New York: Routledge, 2003).

Chapter 1: Geography and vision

1 Stephen Daniels, *Fields of Vision: Landscape and National Identity in England and the United States* (Princeton: Princeton University Press, 1993); Denis Cosgrove, *Social Formation and Symbolic Landscape* (London: Croom Helm, 1984); Denis Cosgrove and Stephen Daniels (eds), *The Iconography of Landscape* (Cambridge: Cambridge University Press, 1988); Kenneth Olwig, *Landscape, Nature and the Body Politic: From Britain's Renaissance to America's New World* (Madison, WI: University of Wisconsin Press, 2002); Gillian Rose, *Feminism and Geography: The Limits of Geographical Knowledge* (Cambridge: Polity Press, 1993), etc.

2 We use rather more scientific terminology for this insight today: 'chaos', and elevate it to the status of a theory.

3 An alternative view of the cosmos is inherently chance-driven: the outcome of the random association and disassociation of atoms is found in Lucretius, especially *De Rerum Naturae* and the Epicurean philosophic tradition more generally. See Chapter 3.

4 Simon Girault, *Globe du monde, contenant un bref traité du ciel et de la Terre* (Lengres: J. des Preyz, 1592). The Renaissance cosmographic tradition that Girault's work represents is discussed by S. K. Heninger Jr in *The Cosmographical Glass: Renaissance Diagrams of the Universe* (San Marino, CA: Huntington Library, 1977).

5 See Cosgrove's chapter 'Cosmographic mapping' in David Woodward (ed.), *The History of Cartography*, vol. III, *Cartography in the European Renaissance* (Chicago: University of Chicago Press, 2007).

6 Svetlana Alpers, *The Art of Describing: Dutch Seventeenth-century Painting* (Chicago: University of Chicago Press, 1983).

7 Images are reproduced in Cosgrove, *History of Cartography*. On Schedel, see Ellen Shaffer, *The Nuremberg Chronicle, a Pictorial World History from the Creation to 1493: A Monograph* (Los Angeles: Dawson's Bookshop, 1950). On Holanda, see Sylvie Deswarte, 'Les De Aetatibus Mundi Imagines de Francisco de Holanda', in

Fondation Eugène Piot, *Monuments et memoires publiés par L'Academie des Inscriptions et Belles Lettres* (Paris, 1983).

8 See Chapter 2 where I argue for a reincorporation of cosmographic concerns into cultural geography.

9 See for example the introductory sections to H. W. Van Loon, *The Home of Mankind: The Story of the World We Live in* (London: G. G. Harrap & Co., [1933] 1946). This enormously popular text was initially published as *Van Loon's Geography* in 1932 and sold 800,000 copies in that year alone before running through ten editions by 1955.

10 James S. Romm, *The Edges of the Earth in Ancient Thought: Geography, Exploration and Fiction* (Princeton: Princeton University Press, 1992); Jean François Staszack, *La Géographie d'avant la géographie: le climat chez Aristote et Hippocrate* (Paris: Harmattan, 1995).

11 J. B. Harley, 'Deconstructing the map', *Cartographica* 26 (1989): 1–20.

12 Christian Jacob, *L'Empire des cartes: approche théorique de la cartographie à travers l'histoire* (Paris: A. Michel, 1992).

13 Lisa Jardine and Jerry Brotton, 'Exchanging identity: breaching the boundaries of Renaissance Europe', in *Global Interests: Renaissance Art between East and West* (Ithaca, NY: Cornell University Press, 2000), pp.11–62.

14 See Chapter 6.

15 'Metageography' (which I use in Chapters 11 and 12) refers to the conception and nomenclature of the larger features of the earth's surface – e.g. continents, oceans, mountain chains etc. See Martin Lewis and Karen Wigen, *The Myth of Continents: A Critique of Metageography* (Berkeley, CA: University of California Press, 1997).

16 Francis Bacon, 'Of Empire', 'Of Plantations', 'Of Gardens', in *Essays* (London: Everyman, 1994), pp.49–52, 88–90, 118–23.

17 Quoted in Victor N. Sholpo, 'The harmony of global space', *Geografity* 1 (1993): 8.

18 Sholpo, 'The harmony of global space', 14.

19 Armand Mattelart, 'Mapping modernity: utopia and communications networks', in Denis Cosgrove, *Mappings* (London: Reaktion Books, 1999), pp.169–92; David Blackbourn, *The Conquest of Nature: Water, Landscape, and the Making of Modern Germany* (New York and London: W. W. Norton, 2006), pp.283–99.

20 Amundsen's locating of flags around the pole to be sure of it, and continued questioning of Peary's achievement, reflect the difficulties of actually locating the pole with precision. On the cultural impacts of these polar adventures see Michael F. Robinson, *The Coldest Crucible: Arctic Exploration and American Culture* (Chicago and London: University of Chicago Press, 2006).

21 Anthony Pagden, *European Encounters with the New World: From Renaissance to Romanticism* (New Haven: Yale University Press, 1993).

22 Francesca Fiorani, *The Marvel of Maps: Art, Cartography and Politics in Renaissance Italy* (New Haven: Yale University Press, 2005).

23 See Denis Cosgrove, *Palladian Landscape* (Leicester: Leicester University Press, 1993), pp.167–87, for a fuller discussion of this map.

24 Edward Casey, *Representing Place: Landscape Painting and Maps* (Minneapolis: University of Minnesota Press, 2002), pp.166–70; 228–30.

25 Roland Barthes, *Camera Lucida* (London: Fontana, 1984).

26 David Woodward, 'Maps and the rationalization of geographic space', in Jay A. Levenson (Ed), *Circa 1492: Art in the Age of Exploration* (Washington, DC: National Gallery of Art; New Haven and London: Yale University Press, 1991), pp.83–8.

27 Denis Cosgrove, 'Ptolemy and Vitruvius: spatial representation in the sixteenth-century texts and commentaries', in Antoine Picon and Alessandra Ponte (eds), *Architecture and the Sciences: Exchanging Metaphors* (New York: Princeton University Press, 2003), pp.20–51.

28 Andrea Palladio, *The Four Books of Architecture*, quoted in Cosgrove, *The Palladian Landscape*, pp.227–8.

29 Christopher Wood, *Albrecht Altdorfer and the Origins of Landscape* (Chicago: University of Chicago Press, 1993); Dagmar Eichberger and Charles Zika (eds), *Dürer and His Culture* (New York: Cambridge University Press, 1998). I discuss this in more detail in Chapter 2.

30 Walter S. Gibson, *'Mirror of the Earth': The World Landscape in Sixteenth-century Flemish Painting* (Princeton: Princeton University Press, 1993).

31 See John Ruskin's Preface to the second edition of *Modern Painters* (London: John Allen, 1844), in *Works of Ruskin*, 3:38. (All references to Ruskin's writings are taken unless otherwise stated from E. T. Cook and Alexander Wedderburn (eds), *The Works of Ruskin*, Library Edition, 39 vols, London: George Allen, 1903–12) and given by volume number and page.) See Denis Cosgrove, 'John Ruskin and the geographical imagination', *Geographical Review* 69 (1979): 43–62.

32 This was the application to visible creation doctrine of 'types' that had originated in the early church fathers' (especially Augustine's) reading of the Hebrew scriptures as an anticipatory narrative of Christ's redemptive mission as detailed in the New Testament. A similar idea underlies the 'theory' of 'intelligent design' currently fashionable among Christian evangelicals unwilling (like Ruskin himself) to accept the implications of Darwinism.

33 John Ruskin, *The Elements of Drawing* (London: Smith, Elder & Co., 1857), pp. xi–xii; Valentin Haecker, *Goethe's morphologische Arbeiten und die neuere Forschung* (Jena: Gustav Fischer, 1927).

34 Today, the solar system.

35 John Ruskin, quoted in Denis Cosgrove and John Thornes, 'Of truth of clouds: John Ruskin and the moral order in landscape', in Douglas Pocock (ed.), *Humanistic Geography and Literature: Essays on the Experience of Place* (London: Croom Helm, 1981), pp.20–46.

36 John Ruskin, *The Storm Cloud of the Nineteenth Century: Two Lectures Delivered at the London Institution, February 4 and 11, 1884*, (Orpington: George Allen, 1884). See also Chapter 7.

37 I owe many of the insights in this section to the work of David Matless whose writings have explored the themes of vision and order in the geographical culture of interwar Britain: David Matless, *Landscape and Englishness* (London: Reaktion, 1998). Barbara Roscoe and Francesco Vallerani's work within the EC funded research project: 'Nature, environment, landscape: European attitudes and discourses in the modern period (1920–1970) with particular attention to water regulation' (EV5V-CT92-0151), interim and final reports 1994, 1995, is also gratefully acknowledged.

38 See Hayden Lorimer, 'Telling small stories: spaces of knowledge and the

practice of geography', *Transactions of the Institute of British Geographers* 28 (2003): 197–217 and 'The geographical field course as active archive', *Cultural Geographies* 10 (2003): 278–308; Sidney William Wooldridge, *The Geographer as Scientist: Essays on the Scope and Nature of Geography* (New York: Greenwood Press, 1969); Sidney William Wooldridge and Geoffrey Hutchings, *London's Countryside: Geographical Fieldwork for Students and Teachers of Geography* (London: Methuen, 1957); Sidney William Wooldridge and Ralph Sisk Morgan, *An Outline of Geomorphology: The Physical Basis of Geography* (London: Longman, 1959); Carl Sauer, *The Morphology of Landscape* (Berkeley, CA: University of California Press, [1925] 1938).

39 Simon Rycroft and Denis Cosgrove, 'Mapping the modern nation: Dudley Stamp and the National Land Use Survey', *History Workshop Journal* 40 (1995): 91–105.

40 William George Hoskins, *The Making of the English Landscape* (London: Hodder and Stoughton, 1965).

41 See the essays in Volker M. Welter and James Lawson, eds. *The City after Patrick Geddes* (Oxford and New York: Peter Lang, 2000); David Matless, 'A modern stream: landscape, modernism and geography', *Environment and Planning D: Society and Space* 10 (1992): 569–88; David Gilbert, David Matless and Brian Short (eds), *Geographies of British Modernity: Space and Society in the Twentieth Century* (Oxford: Blackwell, 2003); Matless, *Landscape and Englishness*; Blackbourn, *The Conquest of Nature*, pp.183–220.

42 Gregory Keypes, *The New Landscape in Art and Science* (Chicago: University of Chicago Press, 1956).

43 Quoted on p.7 in Simon Rycroft, 'The project of modernism in Europe (1920–1970)' in the interim report of the EC project referred to in note 37 above.

44 G. A. Jellicoe, *Studies in Landscape Design*, vol. III (Oxford: Oxford University Press 1966), p.14; quoted in Rycroft, 'The project of modernism', 11.

45 These images are discussed in Denis Cosgrove, 'Contested global visions: One-World, Whole-Earth, and the Apollo space photographs', *Annals of the Association of American Geographers* 84 (1994): 270–94.

46 James Lovelock, *Gaia: A New Look at Life on Earth* (Oxford: Oxford University Press, 1987); Jon Turney, *Lovelock and Gaia: Signs of Life* (New York: Columbia University Press, 2004).

Chapter 2: Extra-terrestrial geography

1 Alexander von Humboldt, *Cosmos: A Sketch of the Physical Description of the Universe*, trans. E. Otté, 2 vols. (Baltimore and London: Johns Hopkins University Press, [1858] 1997).

2 Ann Godlewska, *Geography Unbound* (Chicago and London: University of Chicago Press, 1999); David Livingstone and Charles Withers, *Geography and Enlightenment* (Chicago and London: University of Chicago Press, 1999).

3 There is some evidence of increasing attention among geographers to the cultural geography of planetary space: e.g. Maria D. Lane, 'Geographers of Mars: cartographic inscription and exploration narrative in late Victorian

representations of the Red Planet', *Isis* 96 (2005): 477–506; Maria D. Lane, 'Mapping the Mars canal mania: cartographic projection and the creation of a popular icon', *Imago Mundi* 58/2 (2006): 198–211.

4 These are headings used by von Humboldt in volume II of *Cosmos*.

5 Von Humboldt, *Cosmos*, I, 71.

6 On natural theology, see David N. Livingstone, *The Geographical Tradition* (Oxford: Blackwell, 1992), pp.119–25. Ruskin knew and admired von Humboldt's work but never accepted the German scientist's belief in wholly natural causes of physical phenomena.

7 Von Humboldt, *Cosmos*, I, 75.

8 David Seamon and Arthur Zajonc, *Goethe's Way of Science: A Phenomenology of Nature* (Albany, NY: State University of New York Press, 1998).

9 Von Humboldt, *Cosmos*, I, 78.

10 M. R. Wright, *Cosmology in Antiquity* (New York: Routledge, 1995).

11 Jean François Staszack, *La Géographie d'avant la Géographie: le Climat chez Aristote et Hippocrate* (Paris: Harmattan, 1995).

12 Edward Grant, *Planets, Stars, and Orbs: The Medieval Cosmos, 1200–1687* (Cambridge: Cambridge University Press, 1994); Yi-Fu Tuan, *The Hydrologic Cycle and the Wisdom of God: A Theme in Geoteleology* (Toronto: University of Toronto Press, 1968). See also Yi-Fu Tuan, 'Mountains, ruins and the sentiment of melancholy', *Landscape* 14 (1964): 27–30; Clarence Glacken, *Traces on the Rhodian Shore: Nature and Culture in Western Thought from Ancient Times to the End of the Eighteenth Century* (Berkeley and Los Angeles: University of California Press, 1967); Marjorie Hope Nicholson, *Mountain Gloom and Mountain Glory: The Development of the Aesthetics of the Infinite* (Seattle: University of Washington Press, [1959] 1997).

13 Giorgio Mangani, *Cartografia morale: geografia, persuasione, identità* (Modena: Franco Cosimo Panini, 2006), pp.40–49.

14 <http://psychclassics.yorku.ca/Plato/Timaeus/timaeus3.htm,   paragraph 38>.

15 Ann Blair, *The Theater of Nature: Jean Bodin and Renaissance Science* (Princeton: Princeton University Press, 1997).

16 See Denis Cosgrove, *Apollo's Eye: A Cartographic Genealogy of the Earth in the Western Imagination* (Baltimore: Johns Hopkins University Press, 2001), pp.57–78 for a more detailed discussion of Christ–Apollo.

17 For a detailed discussion of this theory of signs, see Alessandro Scafi, *Mapping Paradise: A History of Heaven on Earth* (London: British Library, 2006).

18 Lesley B. Cormack, *Charting an Empire: Geography at the English Universities, 1580–1620* (Chicago: University of Chicago Press, 1997).

19 Von Humboldt, *Cosmos*, II, 19.

20 Von Humboldt, *Cosmos*, I, 47–54.

21 Denis Cosgrove, 'Images of Renaissance cosmography, 1450–1650', in David Woodward (ed.), *The History of Cartography*, vol. III, *Cartography in the European Renaissance* (Chicago: University of Chicago Press, 2007).

22 Von Humboldt, *Cosmos*, II, 314.

23 Eileen Reeves, *Painting the Heavens: Art and Science in the Age of Galileo* (Princeton: Princeton University Press, 1997).

24 Fernand Hallyn, *The Poetic Structure of the World: Copernicus and Kepler*, trans. Donald

Leslie (New York: Zone Books, 1993), 185–202; S. K. Heninger, *Touches of Sweet Harmony: Pythagorean Cosmology and Renaissance Poetics* (San Marino, CA: Huntington Library, 1974).

25 Cosgrove, 'Renaissance cosmography'.

26 Francesco Berlinghieri, *Geographia* (Florence: Niccolò Todescho, 1482).

27 Brian Vickers (ed.), *Occult and Scientific Mentalities in the Renaissance* (Cambridge: Cambridge University Press, 1984).

28 Sebastian Münster, *Cosmographia* (Basel, 1544); André Thevet, *Cosmographie universelle* (Paris: Guillaume Chandiere, 1575).

29 Frank Lestringant, *Mapping the Renaissance World: The Geographical Imagination in the Age of Discovery*, trans. David Fausett (Berkeley and Los Angeles: University of California Press, 1994), p.130.

30 Jonathan Crary, *Techniques of the Observer: On Vision and Modernity in the Nineteenth Century* (Cambridge, MA: MIT Press, 1991).

31 Walter S. Gibson, *'Mirror of the Earth': The World Landscape in Sixteenth-century Flemish Painting* (Princeton: Princeton University Press, 1989); Christopher Wood, *Albrecht Altdorfer and the Origins of Landscape* (Chicago: University of Chicago Press, 1993).

32 Quoted in Wood, *Albrecht Altdorfer*, p.46.

33 Stephen Hawking, *A Brief History of Time* (Toronto: Bantam, 1988).

34 Wright, *Cosmology in Antiquity*, p.1.

35 Wright, *Cosmology in Antiquity*, p.2.

36 Quoted in Wright, *Cosmology in Antiquity*, p.47.

37 Malins' deep space images are available for viewing at <ftp://ftp.seds.org/pub/images/deepspace/AAT/aat007.gif>

38 Martin Heidegger, 'Building, dwelling, thinking', in David Farrell Krell (ed.), *Martin Heidegger: Basic Writings from 'Being and Time' to 'The Task of Thinking'* (London: Routledge and Kegan Paul, 1978), pp.319–40.

39 Denis Cosgrove, 'Contested global visions: One-World, Whole-Earth, and the Apollo space photographs', *Annals of the Association of American Geographers* 84 (1994): 270–94.

40 Michael Light, *Full Moon* (New York: Alfred Knopf, 1999). The idea of the uncanny in landscape has received considerable attention in recent years. See, for example, Steve Pile, *Real Cities: Modernity, Space and the Phantasmagorias of City Life* (London: Sage, 2005); Anthony Vidler, *The Architectural Uncanny: Essays in the Modern Unhomely* (Cambridge, MA and London: MIT Press, 1992). The uncanniness of extra-terrestrial landscape is a key motif in science fiction and merits more detailed study. See William Fox, *Playa Works: The Myth of the Empty* (Reno: University of Nevada Press, 2002).

41 Light, *Full Moon*, p.3.

42 Yi-Fu Tuan, *Cosmos and Hearth: A Cosmopolite's Viewpoint* (Minneapolis: University of Minnesota Press, 1999).

43 See the special issue of *Geoforum* 'Enacting geographies', 33 (2002); John Wylie, 'Becoming icy: Scott and Amundsen's South Polar voyages, 1910–1913', *Cultural Geographies* 9 (2002): 249–65; John Wylie, 'A single day's walking: narrating self and landscape on the South West Coast Path', *Transactions of the Institute of British Geographers* 30 (2005): 234–47; Bronislaw Szerszynski, Wallace Heim and Claire Waterton (eds), *Nature Performed: Environment, Culture and Performance*

(Oxford: Blackwell, 2004). Neil Lewis, 'The climbing body, nature and the experience of modernity', *Body and Society* 6 (2000): 58–80.

44 Von Humboldt, *Cosmos*, I, 80.

Chapter 3: Gardening the Renaissance world

1 Simon Schama, *Landscape and Memory* (New York: Alfred A. Knopf, 1995).
2 Thomas More, *A Fruitful, Pleasant and Witty Work Called Utopia*, trans. Ralph Robinson (Menston: Scolar Press [1556] 1970).
3 Richard Grove, *Green Imperialism: Colonial Expansion, Tropical Island Edens and the Origins of Environmentalism, 1600–1860* (Cambridge: Cambridge University Press, 1995); Yi-Fu Tuan, *Dominance and Affection: The Making of Pets* (New Haven: Yale University Press, 1984); Marcus Hall, *Earth Repair: A Transatlantic History of Environmental Restoration* (Charlottesville and London: University of Virginia Press, 2005).
4 Jonathan Sawday, *The Body Emblazoned: Dissection and the Human Body in Renaissance Culture* (London and New York: Routledge, 1995).
5 Lucia Tongiorgi Tomasi and Gretchen A. Hirschauer, *The Flowering of Florence: Botanical Art for the Medici* (Washington, DC: Aldershot, 2002). See also Anthony Grafton and Nancy Sirasi (eds), *Natural Particulars: Nature and the Disciplines in Renaissance Europe* (Cambridge, MA and London: MIT Press, 1999).
6 This idea is perfectly expressed in Pico della Mirandola's great statement of humanist philosophy, *Oration on the Dignity of Man* (1486), in which he figures the human suspended midway between angel and beast, free to choose the direction in which to realize the self in life. Available online at <http://www.wsu.edu:8080/~wldcis/world_civ_reader/world_civ_reader_1/pico.html>.
7 Schama, *Landscape and Memory*.
8 See Titian's famous masterpiece *Sacred and Profane Love* (1514: Borghese Gallery, Vatican).
9 Pietro Bembo, *Gli Asolani*, critical ed. Giorgio Dilemmi (Firenze: Accademia della Crusca, 1991).
10 Gianbattista Ramusio, *Navigationi et viaggi* (Venice, 1563–1606).
11 Peter Hulme, *Colonial Encounters: Europe and the Native Caribbean, 1492–1797* (New York: Methuen, 1986).
12 Girolamo Fracastoro, *Fracastoro's Syphilis: Introduction, Text, Translation and Notes*, ARCA, Classical and Medieval Texts, Papers and Monographs 12 (Liverpool: Frances Cairns Publications, 1984), p.84.
13 Francis Bacon, 'On plantations', in *Essays* (London: Everyman, 1994), p.88.
14 Ibid., p.90.
15 Bacon, 'Of empire', 'Of plantations', 'Of gardens' (in *Essays*) all make reference to England's Virginia colonization efforts.
16 Daniele Barbaro, *Lettere*, quoted in I. J. Reist, 'Divine Love and Veronese's frescoes at the Villa Barbaro', *Art Bulletin* 67 (1985): 632.
17 Pollio Vitruvius (ed. D. Barbaro), *I dieci libri dell'architettura di M. Vitruvio: tradutti et commentati da Monsignor Barbaro, eletto patriarca d'Aquileggia, con due tauole, l'una di tutto quelo si contiene per i capi nell'Opera, l'altra per dechiaratione di tutte le cose d'importanza* (Venice: Francesco Marcolini, 1556).
18 Ibid.

19 Denis Cosgrove, *Palladian Landscape* (Leicester: Leicester University Press, 1993), pp.109–10.

20 Nicolò Zeno, *Il sommario, retratto di Monselice* (6.8.1557).

21 This description echoes the long ekphrastic tradition of the *loci amoeni*, dating back to Homer (see especially his description of the Garden of Alkinoos in book VII of the *Odyssey*) and continuing in the Byzantine period. See Liz James and Ruth Webb, 'To understand ultimate things and enter secret places: ekphrasis and art in Byzantium', *Art History* 14 (1991): 1–17.

22 Pietro Bembo, *Petri Bembi Cardinalis Historiae Venetae*, libri XII (Venice: Apud Alid filios, 1551).

23 Schama, *Landscape and Memory*, pp.254–306.

24 These four rivers are personified by Bernini in his great fountain in the Piazza Navona in Rome. The identity of the four rivers said to flow out of the terrestrial paradise was disputed, but commonly they were identified as the Euphrates, Tigris, Ganges and Nile. See Alessandro Scafi, *Mapping Paradise: A History of Heaven on Earth* (London: British Library, 2006).

25 Abraham Ortelius, *Theatrum Orbis Terrarum Parergon; sive Veteris Geographiæ tabulæ, commentariis geographicis et historicis illustratæ. Editio novissima, tabulis aliquot aucta et varie emendata atque innovata, cura e studio Balthasaris Moreti* (Antwerp: Ex officina Plantiniana, Balthsaris Moreti, 1624).

26 Pietro Bembo, *De Aetna ad Angelum Chabrielem Liber* (Venice: In aedibus Aldi, 1495).

27 Hoefnagel's images of his tour with Ortelius were published in Georg Braun and Franz Hogenberg's collection *Civitates Orbis Terrarum* (6 vols., Cologne, 1572–1617). See also Lucia Nuti, 'Alle origini del *Grand Tour*: immagini e cultura della città italiana negli atlanti e nelle cosmografie del secolo XVI', in Giorgio Botta (ed.), *Cultura del viaggio: ricostruzione storico-geografica del territorio* (Milan: Unicopli, 1989), pp.209–52.

28 Ortelius, *Theatrum Orbis Terrarum Parergon*, n.p.

29 Patricia Fortini Brown, *Venice and Antiquity: The Venetian Sense of the Past* (New Haven: Yale University Press, 1999).

30 Barbara E. Mundy, 'Mapping the Aztec capital: the 1524 Nuremberg map of Tenochtitlan, its sources and meanings', *Imago Mundi* 50 (1998): 11–33.

31 Kirsten A. Seaver, 'Norumbega and Harmonia Mundi in sixteenth-century cartography', *Imago Mundi* 50 (1998): 34–58.

Chapter 4: Mapping Arcadia

1 See the introduction to Denis Cosgrove, *Mappings* (London: Reaktion Books, 1999).

2 Virgil, *The Aeneid*, trans. James H. Mantinband (New York: Frederick Ungar, 1964), bk VIII, v.348, p.173.

3 Jacopo Sannazaro, *Arcadia and Piscatorial Eclogues*, trans. Ralph Nash (Detroit: Wayne State University Press, 1966).

4 Ralph Nash, Introduction to Sannazaro, *Arcadia*, pp.24–5.

5 Edmund Spenser, *The Fairie Queene* (London: Dent, 1987); see also Jean-Honoré Fragonard's *Husken*, 1767, in the Wallace Collection, London.

6 Quoted in Carl O. Sauer, *Sixteenth Century North America: The Land and the People as Seen by Europeans* (Berkeley and Los Angeles: University of California Press, 1971), p.54.

7 Anthony Vidler, *The Architectural Uncanny: Essays in the Modern Unhomely* (Cambridge, MA and London: MIT Press, 1992); David Pinder, 'Ghostly footsteps: voices, memories and walks in the city', *Ecumene* 8 (2001): 1–19; Steve Pile, 'Perpetual returns: vampires and the ever-colonized city', in Ryan Bishop, John Phillips and Wei-Wei Yeo (eds), *Postcolonial Urbanism: Southeast Asian Cities and Global Processes* (London: Routledge, 2003); Dydia DeLyser, 'When less is more: absence and social memory in a California ghost town' in Paul Adams et al. (eds), *Textures of Place: Exploring Humanist Geographies* (Minneapolis: University of Minnesota Press, 2001).

8 Stephanie Barron et al., *Made in California: Art, Image, and Identity 1900–2000* (Los Angeles: Los Angeles County Museum of Art; Berkeley, CA: University of California Press, 2000).

9 Richard A. Walker, 'California's golden road to riches: natural resources and regional capitalism 1848–1940', *Annals of the Association of American Geographers* 91 (2001): 167–99.

10 Minick uses a classic view of the Merced Falls as the prospect point enclosed between the wings of a stage setting (or coulisses) at the depth of receding planes of light and shadow, a composition refined in Claude's landscapes, many of which are set in explicitly Arcadian locations: for example, *Landscape with Ascanius Shooting the Stag of Sylvia* (1688; Ashmolean Museum, Oxford).

11 Dean MacCannell, 'Nature incorporated', in *Empty Meeting Grounds: The Tourist Papers* (London and New York: Routledge, 1992), pp.114–20.

12 John Muir. Despite my best attempts, I have been unable to locate the original source of this statement.

13 Finis Dunaway, *Natural Visions: The Power of Images in American Environmental Reform* (Chicago: University of Chicago Press, 2005).

14 See for example, Tariq Jazeel, 'Nature, nationhood and the poetics of meaning in Ruhuna (Yala) National Park, Sri Lanka', *Cultural Geographies* 12 (2005): 199–228; Roderick P. Neumann, *Imposing Wilderness: Struggles over Livelihood and Nature Preservation in Africa*, California Studies in Critical Human Geography 4 (Berkeley, Los Angeles, London: University of California Press, 1998); Marcus Hall, *Earth Repair: A Transatlantic History of Environmental Restoration* (Charlottesburg and London: University of Virginia Press, 2005).

15 Hall, *Earth Repair*, argues that return to a pristine, wilderness state is the controlling vision of American environmentalism, while that of gardening nature characterizes European attitudes.

16 William Cronon, *Uncommon Ground: Rethinking the Human Place in Nature* (New York: W. W. Norton, 1996); Sarah Whatmore, *Hybrid Geographies: Nature, Cultures, Spaces* (London and Thousand Oaks, CA: Sage, 2002); Bronislaw Szerszynski, Wallace Heim and Claire Waterton (eds), *Nature Performed: Environment, Culture and Performance* (Oxford: Blackwell, 2003).

17 I pose this as a broader question than simply what it meant to designate Yosemite as a park. That question has been broached by Ken Olwig, whose discussion of Olmstead's ideas relates its 'emparkment' to the aesthetics of

urban open space, and thus back to common lands and spaces in feudalism rather than the aristocratic spaces of the hunting parks. See also the discussion in Hall, *Earth Repair*, pp.139–46.

18 Sannazaro, *Arcadia*, p.31.

19 Ibid., Eclogue I, pp.32–3.

20 Ibid., p.101 passim.

21 Erwin Panofsky, 'Et in *Arcadia Ego*: Or the conception of transcience in Poussin and Watteau', in *Philosophy and History: Essays Presented to Ernest Cassirer*, R. Klibansky and H. J. Paton (eds) (Oxford: Clarendon Press, 1936).

22 Sannazaro, *Arcadia*, 78.

23 Richard Jenkyns, *Virgil's Experience: Nature and History* (Oxford: Clarendon Press, 1998).

24 Virgil, Eclogue vii. Virgil's Eclogues are available online at <http://classics. mit.edu/Virgil/eclogue.html>

25 Jenkyns, *Virgil's Experience*, p.177.

26 Virgil, *Aeneid*, bk VIII, v.351–354, p.173.

27 Ibid., bk VIII, v.360–361, p.173.

28 See Joseph Rykwert, *On Adam's House in Paradise: The Idea of the Primitive Hut in Architectural Theory* (New York: Museum of Art, 1972); see also the series of paintings on the origins of human community by Piero di Cosimo, especially *The Forest Fire*, c.1505 (Ashmolean Museum, Oxford), *Vulcan and Aeolus*, c.1490 (National Gallery of Canada) and *The Discovery of Honey by Bacchus*, c.1499 (Worcester Art Museum, Massachusetts).

29 Virgil *Aeneid*, bk VIII, v.314–18, p.172.

30 Lucretius, *On the Nature of the Universe: De Rerum Natura* (Harmondsworth: Penguin Books, 1951, 1967), trans. R. E. Latham, Book 1.

31 Op. cit., Book 5.

32 Op. cit., Book 5.

33 Op. cit., Book 5.

34 Op. cit., Book 5.

35 See the introduction to Diana Nemiroff, *Elusive Paradise: The Millennium Prize* (Ottawa: National Gallery of Canada, 2001).

36 The full text of the Act is available at: <http://www.ourdocuments.gov/ doc.php?doc=45&page=transcript>

Chapter 5: Measures of America

1 On John Winthrop, first governor of Massachusetts Colony, see Loren Baritz, 'The idea of the West', *American Historical Review* 66 (1961): 618–40; Hector St John de Crèvecoeur, *Letters from an American Farmer*, Letter 3, 1782, available at <http://xroads.virginia.edu/~hyper/CREV/contents.html>; Jean Baudrillard, *America*, trans. Chris Turner (London and New York: Verso, 1989).

2 Frederick Jackson Turner, *The Frontier in American History* (Tucson: University of Arizona Press, [1893] 1986).

3 The original version of this essay appears in James Corner and Alex MacLean, *Taking Measures: Across the American Landscape* (New Haven: Yale University Press,

1996) which reproduces the aerial landscape photographs of Alex MacLean whose images perfectly capture the logic of American landscape.

4 Robert Wohl, *A Passion for Wings: Aviation and the Western Imagination, 1908–1918* (New Haven: Yale University Press, 1994); and *The Spectacle of Flight: Aviation and the Western Imagination, 1920–1950* (New Haven: Yale University Press, 2005).

5 Stephen Kern, *The Culture of Space and Time, 1880–1918* (Cambridge, MA: Harvard University Press, 1983). See the concept of space-time compression in David Harvey, *The Condition of Postmodernity: An Enquiry into the Origins of Cultural Change* (Oxford and New York: Blackwell, 1989).

6 Wohl, *Spectacle of Flight*, pp.37–8.

7 This vision was widely promoted among artists and architects, especially Italian Futurists: see Bruno Mantura, Patrizia Rosazza-Ferraris and Livia Velani (eds), *Futurism in Flight: 'Aeropittura' Paintings and Sculptures of Man's Conquest of Space* (1913–1945) (Rome: De Luca Edizioni d'Arte, 1990). See also the images of aircraft and architecture in Le Corbusier's *Vers une Architecture* (Paris: n.p., 1923).

8 Finis Dunaway, *Natural Visions: The Power of Images in American Environmental Reform* (Chicago: University of Chicago Press, 2005).

9 Erwin Anton Gutkind, 'Our world from the air: conflict and adaptation', in William L. Thomas (ed.), *Man's Role in Changing the Face of the Earth* (Chicago: University of Chicago Press, 1956), p.1.

10 'Whole and unshadowed', is not of course technically correct: only one hemisphere of the globe is visible. See Denis Cosgrove, 'Contested global visions: *One World, Whole-Earth*, and the Apollo space photographs', *Annals of the Association of American Geographers* 84 (1994): 270–94.

11 Of course, the metageographical terminology of continents, like oceans, is cultural rather than natural; see Martin Lewis and Karen Wigen, *The Myth of Continents: A Critique of Metageography* (Berkeley, CA: University of California Press, 1997).

12 Fernand Hallyn, *The Poetic Structure of the World: Copernicus and Kepler* (Cambridge, MA: MIT Press, 1993).

13 Gaetano Ferro, Luisa Faldini and Marica Milanesi, *Columbian Iconography* (Rome: Istituto Poligrafico e Zecca dello Stato, 1992).

14 Pico della Mirandola, *Oration on the dignity of man* (1486), available online at <http://www.wsu.edu:8080/~wldciv/world_civ_reader/world_civ-_reader_1/pico.html>

15 Quoted in William Boelhower, *Through a Glass Darkly: Ethnic Semiosis in American Literature* (Oxford: Oxford University Press, 1985), pp.50–1.

16 G. Malcolm Lewis, 'Maps, mapmaking, and map use by native North Americans', in David Woodward and G. Malcolm Lewis (eds), *The History of Cartography*, vol. II, book 3, *Cartography in the Traditional African, American, Arctic, Australian, and Pacific Societies* (Chicago: University of Chicago Press, 1998), p.110.

17 Baritz, 'Idea of the West'.

18 On the debate between Buffon and Jefferson on American nature see Paul Semonin, *American Monster: How the Nation's First Prehistoric Creature Became a Symbol of National Identity* (New York: New York University Press, 2000).

19 Robert Lawson Peebles, *Landscape and Written Expression in Revolutionary America: The World Turned Upside Down* (Cambridge: Cambridge University Press, 1988).

20 Barbara Novak, Nature and Culture: American Landscape and Painting 1825–1875 (New York: Oxford University Press, 1980).

21 Archibald MacLeish, 'America Was Promises', in Collected Poems 1917–1952 (Boston: Houghton Mifflin, 1962).

22 David Matless, 'A modern stream: water, landscape, modernism and geography', Environment and Planning D: Society and Space 10 (1992): 569–88.

23 David E. Nye, America as Second Creation: Technology and Narratives of New Beginnings (Cambridge, MA: MIT Press, 2003); see also Dunaway, Natural Visions.

24 Carl Sauer talks about the morphologic eye as a prerequisite for geographical scholarship in 'Foreword to historical geography', in John Leighly (ed.), Land and Life: Selections from the Writings of Carl Ortwin Sauer (Berkeley and Los Angeles: University of California Press, 1963); see also Dunaway, Natural Visions, pp.63–81.

25 Nye, America as Second Creation.

26 Thomas Stearns Eliot, The Waste Land (New York: Boni and Liveright, 1922).

27 Scott Kirsch, Proving Grounds: Project Plowshare and the Unrealized Dream of Nuclear Earthmoving (New Brunswick, NJ and London: Rutgers University Press, 2005).

28 See my discussion of these attitudes in Chapter 11.

29 Richard Walker, 'California's golden road to riches: natural resources and regional capitalism 1848–1940', Annals of the Association of American Geographers 91 (2001): 167–99.

30 Pico, Oration on the dignity of man (1486).

Chapter 6: Wilderness, habitable earth and the nation

1 Henry David Thoreau, 'Walking', in Civil Disobedience, and Other Essays (New York: Dover Publications; London: Constable, [1862] 1993).

2 Max Oelschlaeger, The Idea of Wilderness, from Prehistory to the Age of Ecology (New Haven and London: Yale University Press, 1991).

3 Yi-Fu Tuan made this observation at a general level 30 years ago in his Topophilia (Englewood Cliffs, NJ: Prentice Hall, 1974).

4 For the argument to justify this claim, see Denis Cosgrove's Apollo's Eye: A Cartographic Genealogy of the Earth in the Western Imagination (Baltimore: Johns Hopkins University Press, 2001).

5 Alfred Crosby, Ecological Imperialism: The Biological Expansion of Europe, 900–1900 (Cambridge: Cambridge University Press, 1986); David Woodward, 'Medieval mappae mundi', in Brian Harley and David Woodward (eds), The History of Cartography vol. I, Cartography in Prehistoric, Ancient, and Medieval Europe and the Mediterranean (Chicago and London: University of Chicago Press, 1987), pp.286–368.

6 Natalia Lozovsky, The Earth Is Our Book: Geographical Knowledge in the Latin West ca. 400–1100 (Chicago: University of Michigan Press, 2000), esp. pp.139–55.

7 John Block Friedman, The Monstrous Races in Medieval Art and Thought (Syracuse, NY: Syracuse University Press, 2000).

8 Oelschlaeger prefers to see this complex of ideas as an atavistic social memory from the neolithic, and contrasts it to a paleolithic 'idea of wilderness'. Such an interpretation is understandable in view of the biological interpretation of humanity upon which his thesis rests.

9 The cycle of civilization, progressing through 'ages' of gold, silver, bronze and iron, is a continuing thread in the classical literary tradition, passing from Hesiod's *Works and Days* through Ovid's *Metamorphoses* and Virgil's triad of *Eclogues, Georgics and Aeneid*. See, among others, Clarence Glacken, *Traces on the Rhodian Shore: Nature and Culture in Western Thought from Ancient Times to the End of the Eighteenth Century* (Berkeley: California University Press, 1967); and Samuel J. Edgerton, 'From mappa mundi to mental matrix to Christian empire', in David Woodward (ed.), *Art and Cartography: Six Essays* (Chicago: Chicago University Press, 1987), pp. 10–50.

10 Northrop Frye, *The Great Code: The Bible and Literature* (San Diego: Harcourt Brace, 1982).

11 Loren Baritz, 'The idea of the West', *American Historical Review* 66 (1961): 618–40; John Huxtable Elliott, *The Old World and the New 1492–1650* (Cambridge: Cambridge University Press, 1992), p.94. The idea that temporal history is a succession of empires and cultures following the sun's westerly path (*translatio imperii*) can be traced back to Augustine. Alessandro Scafi, *Mapping Paradise: A History of Heaven on Earth* (London: British Library, 2006), pp. 126–8.

12 Denis Cosgrove, 'The picturesque city: nature, nations and the urban since the eighteenth century', in T. M. Kristensen et al. (eds), *City and Nature: Changing Relations in Time and Space* (Odense: Odense University Press, Denmark, 1993), pp. 44–58.

13 The history and historiography of encounter and its cultural effects have been wholly rethought since the 500-year Columbian anniversary in 1992. See especially Elliott, *Old World and the New*; Anthony Pagden, *European Encounters with the New World: From Renaissance to Romanticism* (New Haven: Yale University Press, 1993); Eviatar Zerubavel, *Terra Cognita: The Mental Discovery of America* (New Brunswick, NJ: Rutgers University Press, 1992); and the collection of essays and illustrations in Jay A. Levenson (ed.), *Circa 1492: Art in the Age of Exploration* (Washington, New Haven and London: National Gallery of Art/Yale University Press, 1991).

14 Hartmann Schedel was a publisher whose 1493 *Chronicle of the World* was structured as a universal history, and whose contributors and illustrators, including Albrech Dürer, were drawn from the humanist community in the great merchant city.

15 Peter Hulme, *Colonial Encounters: Europe and the Native Caribbean, 1492–1797* (New York: Methuen, 1986).

16 Ibid., p.83.

17 Both these nineteenth-century expeditions (one into the northern polar seas in search of the Northwest Passage, the other across the desert west of the United States) met with severe winter weather that isolated and condemned every member. In the Franklin case it appears that lead poisoning from canned meat was at least as responsible as starvation and exposure for the failure. The tragedies are, however, etched into historical memory by tales of their survivors resorting to cannibalism; profoundly shocking to the self-image of Europeans and Americans in an era of scientific racism that placed them at the pinnacle of a human hierarchy, and in the years of intense debate over our animal nature provoked by publication of Darwin's *Origin of Species*. For a discussion of the cultural impacts of these events in the USA see Michael F. Robinson, *The Coldest*

Crucible: Arctic Exploration and American Culture (Chicago: University of Chicago Press, 2006).

18   Denis Cosgrove, *Social Formation and Symbolic Landscape* (London: Croom Helm, 1984); Peter G. Rowe, *Making a Middle Landscape* (Cambridge, MA: MIT Press, 1991).

19   See Chapters 7 and 8 on John Ruskin, and David Blackbourn, *The Conquest of Nature: Water, Landscape, and the Making of Modern Germany* (New York: W. W. Norton 2006).

20   Denis Cosgrove, Barbara Roscoe and Simon Rycroft, 'Landscape and identity at Ladybower Reservoir and Rutland water', *Transactions of the Institute of British Geographers* ns 21/3 (1996): 534–51; Peter Taylor, 'The English and their Englishness: a curiously mysterious, elusive and little understood people', *Scottish Geographical Magazine* 107 (1991): 146–61. Significantly, Theodore de Bry's *America* illustrations for John White's account of the English settlement of Virginia make an explicit and direct graphic connection between the aboriginal peoples of the New World and ancient Britons: Michael Alexander, *Discovering The New World: Plates Engraved by Theodore de Bry* (New York: Harper and Row, 1976).

21   Simon Schama, *Landscape and Memory* (New York: Alfred A. Kropf, 1995).

22   The literature on the national park idea is vast, and in the past decade has taken on an increasingly critical edge. See Roderick Nash, *Wilderness and the American Mind* (New Haven and London: Yale University Press, 1973); Donald Worster, *Nature's Economy: A History of Ecological Ideas* (Cambridge: Cambridge University Press, 1977); John MacKenzie. *Empire of Nature: Hunting, Conservation, and British Imperialism* (Manchester: Manchester University Press, 1988); Roderick P. Neumann, *Imposing Wilderness: Struggles over Livelihood and Nature Preservation in Africa*, California Studies in Critical Human Geography 4 (Berkeley, Los Angeles, London: University of California Press, 1998); Marcus Hall, *Earth Repair: A Transatlantic History of Environmental Restoration* (Charlottesville and London: University of Virginia Press, 2005). On the role of photography in shaping ideas of wilderness in the post-war years, see Finis Dunaway, *Natural Visions: The Power of Images in Amercian Environmental Reform* (Chicago: University of Chicago Press, 2005); see also Steven Hoelscher, 'The photographic construction of tourist space in Victorian America', *Geographical Review* 88 (1998): 548–69.

23   Neumann, *Imposing Wilderness*.

24   The Act of Congress is quoted in Chapter 4. For the full text see <www.ourdocuments.gov/doc.php?doc=45&page=transcript>

25   Schama, *Landscape and Memory*; David Livingstone, *Nathaniel Southgate Shaler and the Culture of American Science* (Tuscaloosa: University of Alabama Press, 1987).

26   Turner's essay drew explicitly on the newly elaborated germ theory of disease. A similar approach was used also by early twentieth-century Viennese art historians, such as Joseph Strygowski; see Christopher Wood, *The Vienna School Reader: Politics and Art Historical Method in the 1930s* (New York: Zone Books, 2000).

27   Gerry Kearns, 'Closed space and political practice: Frederick Jackson Turner and Halford Mackinder', *Environment and Planning D: Society and Space* 2 (1984): 23–34.

28   Barbara Novak, *Nature and Culture: American Landscape and Painting 1825–1875* (New York: Oxford University Press, 1980); Stephen Daniels, *Fields of Vision: Landscape and National Identity in England and the United States* (Princeton: Princeton University Press, 1993).

29 Maldwin Allen Jones, *American Immigration* (Chicago: University of Chicago Press, 1960).

30 On the American impacts of Darwinian thought and scientific racism, see David N. Livingstone, *Darwin's Forgotten Defenders: The Encounter Between Evangelical Theology and Evolutionary Thought* (Edinburgh: Scottish Academic Press, 1987).

31 Kevin Starr, *Americans and the California Dream 1850–1915* (New York: Oxford University Press, 1973).

32 David N. Livingstone, *The Geographical Tradition: Episodes in the History of a Contested Enterprise* (Oxford: Blackwell, 1992), pp.206–8.

33 New York politician Prescott F. Hall, 1894, quoted in Jones, *American Immigration*, p.259.

34 Mark Bassin, 'Turner, Solov'ev, and the "Frontier Hypothesis": the nationalist significance of open spaces', *Journal of Modern History* 65 (1993): 473–511.

35 Terence Young, 'Social reform through parks: the American Civic Association's program for a better America', *Journal of Historical Geography* 22 (1996): 460–72.

36 Kenneth Olwig, 'Sexual cosmology: nation and landscape at the conceptual interstices of nature and culture', in Barbara Bender (ed.), *Landscape: Politics and Perspectives* (London: Berg, 1993).

37 T. J. Jackson Lears, *No Place of Grace: Antimodernism and the Transformation of American Culture 1880–1920* (New York: Pantheon, 1981), p.146.

38 Quoted in Jackson Lears, *No Place of Grace*, p.147.

39 John MacKenzie, *Propaganda and Empire: The Manipulation of British Public Opinion, 1880–1960* (Manchester: Manchester University Press, 1984); Anne McClintock, *Imperial Leather: Race, Gender and Sexuality in the Colonial Context* (London: Routledge, 1995).

40 Gary Willis, 'American Man', *New York Review of Books*, 6 March (1977): 30–3.

41 See Dunaway's discussion of David Brower's responses to criticism of Sierra Club wilderness policies (*Natural Visions*); and the vitriolic debate stimulated by William Cronon's 'The trouble with wilderness; or, getting back to the wrong nature', in Cronon (ed.), *Uncommon Ground: Rethinking the Human Place in Nature* (New York: W. W. Norton, 1996), pp.69–90. The Sierra Club's recent history of internal division over immigration policy is instructive here.

42 Dunaway, *Natural Visions*.

43 Michael Frome, *Battle for the Wilderness* (London: Praeger, 1974), p.13.

44 Joe Hermer, *Regulating Eden: The Order of Nature in North American Parks* (Toronto: University of Toronto Press, 2002).

45 Examples of this large genre of popular wilderness writing include Barry Lopez, *Arctic Dreams: Imagination and Desire in a Northern Landscape* (New York; Scribner, 1986); Peter Matthiessen, *The Snow Leopard* (New York: Viking Press, 1978); and David Pitt-Brooke, *Chasing Clayoquot: A Wilderness Almanac* (Vancouver: Raincoast Books, 2004).

46 The arguments and literature are ably summed up in Hall, *Earth Repair*.

47 Noel Castree and Bruce Braun (eds), *Social Nature: Theory, Practice and Politics* (Oxford: Blackwell, 2001).

48 See, for example, Laura Cameron and David Matless, 'Benign ecology: Marietta Pallis and the floating fen of the delta of the Danube, 1912–16', *Cultural Geographies* 10 (2003): 253–77. Ernst Haeckel, Social Darwinist and one of the fathers

of ecological science, was also one of the earliest to argue that Nature had rights. More recent supporters of bio-regionalism and deep ecology sometimes betray similar attitudes to society as early twentieth-century eugenicists, especially in their views on human population control.

49 Oelschlaeger, *The Idea of Wilderness*.

50 Denis Cosgrove, 'Contested global visions: One World, Whole-Earth, and the Apollo space photographs', *Annals of the Association of American Geographers* 84 (1994): 270–94.

51 Dean MacCannell, 'Nature incorporated' and 'Cannibalism today', in *Empty Meeting Grounds: The Tourist Papers* (London: Routledge, 1992), pp.17–73, 114–20.

52 In 1996 the Sierra Club voted to change its long-standing policy on US population stabilization which implied a restriction on immigration. Thenceforth the Club would be neutral on the question of American population numbers. This caused a significant and continuing polemic within the organization over current US immigration policy, which may be followed on a number of weblogs, for example <http://www.susps.org/index.html>. See also and Alexander Wilson, *The Culture of Nature: North American Landscape from Disney to the Exxon Valdez* (Oxford: Blackwell, 1992).

53 Dunaway, *Natural Visions*; Mark Klett (ed.), *Third Views, Second Sights: A Rephotographic Survey of the American West* (Santa Fe: Museum of New Mexico and the Center for American Places, 2004) and Mark Klett (et al.) *Yosemite in Time: Ice Ages, Tree Clocks, Ghost Rivers* (San Antonio, TX: Trinity University Press, 2005).

54 Mark Klett (et al.) *Yosemite in Time: Ice Ages, Tree Clocks, Ghost Rivers* (San Antonio, TX: Trinity University Press, 2005).

Chapter 7: The morphological eye

1 'Geophil', anonymous manuscript (Oxford: School of Geography archives, Oxford University Centre for the Environment), 1914.

2 John Ruskin, 'The Elements of Drawing and Perspective', in E. T. Cook and A. Wedderburn (eds), *The Works of Ruskin*, Library Edition, 39 vols (London: George Allen, 1903–12), 15:13.

3 Thomas Henry Huxley, *Physiography: An Introduction to the Study of Nature* (London: Macmillan, 1877). See the discussion in David Ross Stoddart, *On Geography and Its History* (Oxford: Blackwell, 1986), pp.180–203.

4 Huxley, *Physiography*, quoted in Stoddart, *On Geography*, p.188.

5 Ibid., p.190.

6 Ruskin, 'The Poetry of Architecture', *Works of Ruskin*, 1:5.

7 Ruskin, 'The Poetry of Architecture', *Works of Ruskin*, 1:74.

8 I outline Ruskin's ideas on landscape and its relations with cultural geography in Denis Cosgrove, 'John Ruskin and the geographical imagination', *Geographical Review* 69 (1979): 43–62.

9 Halford J. Mackinder, 'On the scope and methods of geography' [1887], in *Democratic Ideals and Reality* (New York: W. W. Norton, 1962).

10 J. F. Heyes, 'A plea for geography', *Oxford Magazine* (1886): 8–12. The copy in the School of Geography archives is attached to an annotated manuscript by

Heyes with the words: 'Geosophy: a concept and an ideal in 1887 when resident in Oxford.' This sketches a vision of geography much closer to natural theology.

11 Heyes, 'Geosophy', n.p. Whether the American geographer and historian of cartography, John Kirtland Wright, knew of Heyes's work or drew on his usage in framing his own concept of 'geosophy' is unclear (John Kirtland Wright, 'Terrae incognitae: the place of the imagination in geography', *Annals of the Association of American Geographers* 37, 1947: 1–15). But the term seems to have a recurrent appeal to those concerned with the imaginative dimensions of geographical experience and knowledge.

12 D. Ian Scargill, 'The RGS and the foundation of geography at Oxford', *Geographical Journal* 142 (1976): 438–61.

13 Francis O'Gorman, 'The eagle and the whale? Ruskin's argument with John Tyndall', in Michael Wheeler (ed.), *Time and Tide: Ruskin Studies* (London: Pilkington Press, 1996), pp.45–64, ref p.52. See also John Moore, 'The nexus between Ruskin's mythopoeic science and his architectural thought', in M. J. Ostwald and R. J. Moore (eds), *Re-Framing Architecture: Theory, Science and Myth* (Sydney: Arcadia Press, 2000), pp.143–54.

14 Ruskin, 'Deucalion', *Works of Ruskin*, 26:97.

15 Ruskin, 'Lectures on Landscape', *Works of Ruskin*, 22:14.

16 Ruskin, 'Deucalion', *Works of Ruskin*, 26:102.

17 Ruskin, 'Lectures on Landscape', *Works of Ruskin*, 22:12.

18 Ruskin, 'Deucalion', *Works of Ruskin*, 26:98–9.

19 Ruskin, 'Modern Painters Vol IV', *Works of Ruskin*, 6:475.

20 Stoddart, *On Geography*, 65. Murchison was President of the RGS for three separate periods between 1843 and 1871.

21 Paul Wilson, '"Over yonder are the Andes": Reading Ruskin reading Humboldt', in Michael Wheeler (ed.), *Time and Tide: Ruskin and Science* (London: Pilkington Press, 1996), pp.65–84.

22 In this Ruskin's geographical vision seems closer to Karl Ritter's than to von Humboldt's, but there is no evidence that Ruskin had read Ritter's work or was influenced by his ideas.

23 Ruskin, 'Deucalion', *Works of Ruskin*, 26:339.

24 Ruskin, 'Lectures on Landscape', *Works of Ruskin*, 22:15–16.

25 Ruskin 'The Poetry of Architecture', *Works of Ruskin*, 1:67–71; See Cosgrove, 'Geographical imagination' for a fuller discussion of this landscape classification by colour.

26 Ruskin, 'Stones of Venice', *Works of Ruskin*, 10:185–8. The passage is quoted in full and discussed in detail in Chapter 8.

27 Ruskin, 'The Bible of Amiens', *Works of Ruskin*, 33:172.

28 Ruskin 'Fors Clavigera', *Works of Ruskin*, 29:505.

29 Ruskin, 'Fors Clavigera', *Works of Ruskin*, 29:506. The physiographic maps developed by the Hungarian-American cartographer Erwin Raisz in the 1920s at Harvard to illustrate W. M. Davis's geomorphological theories fulfil these principles remarkably fully.

30 Ibid.

31 Ruskin, 'Lectures on Architecture and Painting', *Works of Ruskin*, 12:5.

32 Robert Hewison, *The Ruskin Art Collection at Oxford: Catalogue of the Rudimentary Series, in the arrangement of 1873, with Ruskin's comments of 1878* (London: Lion and Unicorn Press, 1984). See also Robert Hewison, *Ruskin and Oxford: The Art of Education* (Oxford: Clarendon Press, 1996). The materials are now available online together with notes from Ruskin's lectures at <http://ruskin.oucs.ox.ac.uk/info/methods.php>

33 Ruskin, 'Modern Painters', *Works of Ruskin*, 6:46.

34 Ruskin, 'The Elements of Drawing', *Works of Ruskin*, 15:440.

35 Ruskin, 'The Elements of Drawing', *Works of Ruskin*, 15:445.

36 Outline syllabus for the Diploma Course in Geography, 1899 (School of Geography archives, Oxford University Centre for the Environment), n.p.

37 Outline syllabus for the Diploma Course, n.p.

38 Ruskin, 'The Bible of Amiens', *Works of Ruskin*, 33:88.

Chapter 8: Ruskin's European visions

1 Yuko Kikuchi and Toshio Watanabe, 'Ruskin in Japan, 1890–1940: nature for art, art or life', *Journal of Design History* 18 (2005): 309–11.

2 This chapter originated as a paper delivered at a Victorian Studies conference in 2005, the theme of which was Victorian Europeans. This was a rare and welcome exploration of connections and reciprocal influences that, as Ruskin's case reveals, were much richer than implied by that hoary and apocryphal example of late nineteenth-century British isolationism: the supposed *Times* headline 'Dense fog in Channel, continent isolated'.

3 Giorgio Mangani, *Cartografia Morale: Geografia, Persuasione, Identità* (Modena: Franco Cosimo Panini, 2006), pp.40–3.

4 Martin Lewis and Karen Wigen, *The Myth of Continents: A Critique of Metageography* (Berkeley, CA: University of California Press, 1997).

5 Denys Hay, *Europe: The Emergence of an Idea* (Edinburgh: Edinburgh University Press, 1957); Michael Heffernan, *The Meaning of Europe: Geography and Geopolitics* (New York: Arnold, 1998).

6 The best summary of Ruskin's life is Tim Hilton's two-volume biography *John Ruskin* and *John Ruskin: The Later Years* (New Haven: Yale University Press, [1985] 2000).

7 The full passage is reproduced at <http://art-bin.com/art/oruskin1.html>. Unfortunately the original source is not provided in full and I have so far failed to verify the manuscript or printed copy.

8 Stephen Daniels, *Humphry Repton: Landscape Gardening and the Geography of Georgian England* (New Haven: Yale University Press, 1999); see also Sidney Robinson, *Inquiry into the Picturesque* (Chicago and London: University of Chicago Press, 1991).

9 See note 7 above.

10 Svetlana Boym, *The Future of Nostalgia* (New York: Basic Books, 2002).

11 John Claudius Loudon, *Loudon's Encyclopaedia of Gardening* (London: Longman, Rees, Orme, Brown and Green, 1822); *Loudon's Encyclopaedia of Cottage, Farm and Villa Architecture* (London: n.p., 1833); Isabella Beeton, *Mrs Beeton's Book of Household Management* (London: S. O. Beeton, 1861).

12 Joseph Gluckstein Links, *The Ruskins in Normandy: A Tour in 1858 with Murray's Handbook* (London: Murray, 1968).

13 The itinerary was fairly conventional. See Jeremy Black, *The British Abroad: The Grand Tour in the Eighteenth Century* (Stroud: Alan Sutton, 1992).

14 Ruskin, 'The Poetry of Architecture', *Works of Ruskin*, 1:1.

15 Ruskin, 'The Poetry of Architecture', *Works of Ruskin*, 1.

16 Thomas da Costa Kaufmann, *Towards a Geography of Art* (Chicago and London: University of Chicago Press, 2004), esp p.43–106; David Blackbourn, *The Conquest of Nature: Water, Landscape, and the Making of Modern Germany* (New York: W. W. Norton, 2006).

17 Ruskin, 'The Stones of Venice', *Works of Ruskin*, 10:185–8.

18 Ibid., 9:17.

19 Ibid., 9:356.

20 John Ruskin, *Ruskin's Letters from Venice 1851–52*, ed. John Lewis Bradley (New Haven: Yale University Press, 1995), Letter 34.

21 Ruskin, 'The Stones of Venice', *Works of Ruskin*, 9:28.

22 Ibid., 10:118.

23 John Ruskin, Letter to the *Pall Mall Gazette*, 16 March 1872, in *Works of Ruskin*, 10:459.

24 Ruskin 'The Bible of Amiens', *Works of Ruskin*, 33:88.

25 Ibid., 33:172.

26 Ibid., 33:57–8.

27 Ibid., 33:90. Ruskin's ideas on map and geographical education are more fully treated in Chapter 7.

28 Ibid., 33:99. It is instructive to compare Ruskin's division of the Roman empire with Strabo's in the *Geography*: Daniela Dueck, Hugh Lindsay and Sarah Pothecary, *Strabo's Cultural Geography: The Making of a Kolossourgia* (Cambridge: Cambridge University Press, 2005).

29 Ruskin, 'The Bible of Amiens', *Works of Ruskin*, 33:93.

30 Ibid., 33:91.

31 Ibid., 33:94–5.

32 Dueck et al., *Strabo's Cultural Geography*.

33 Ruskin, 'The Storm Cloud of the Nineteenth Century', *Works of Ruskin*, 34:33.

34 Ibid., 34:33.

35 Kaufmann, *Geography of Art*, p.342.

Chapter 9: Moving maps

1 Among the very large number of examples are David Buisseret, *Monarchs, Ministers, and Maps: The Emergence of Cartography as a Tool of Government in Early Modern Europe* (Chicago: Chicago University Press, 1992); Laura Hostetler, *Qing Colonial Enterprise: Ethnography and Geography in Early Modern China* (Chicago: Chicago University Press, 2001); Graham D. Burnett, *Masters of All They Surveyed: Exploration, Geography, and a British El Dorado* (Chicago: University of Chicago Press, 2000); Tom Koch, *Cartographies of Disease: Maps, Mapping and Medicine* (Redlands, CA: Esri Press, 2005).

2 See the essays in the special issue of *Cartographic Perspectives* 53 (2006).

3 See the introduction to Denis Cosgrove (ed.), *Mappings* (London: Reaktion Books, 1999), pp.1–23.

4 Mark Denil, 'Cartographic design: rhetoric and persuasion', *Cartographic Perspectives* 45 (2003): 8–67.

5 Koch, *Cartographies of Disease*; Keith Clarke and John Cloud, 'On the origins of analytical cartography', *Cartography and Geographic Information Science* 27 (2000): 195.

6 For an early critical discussion see John Pickles, *Ground Truth: The Social Implications of Geographic Information Systems* (New York: Guilford Press, 1995).

7 On the history of the atlas, see David Turnbull, *Maps Are Territories, Science Is an Atlas: A Portfolio of Exhibits* (Chicago: University of Chicago Press, 1993). On Ortelius, see Marcel van den Broecke, Peter van der Krogt and Peter Meurer (eds), *Abraham Ortelius and the First Atlas: Essays Commemorating the Quadricentennial of his Death, 1598–1998* (Utrecht: HES Publishers, 1996); Giorgio Mangani, *Il mondo di Abramo Ortelio: misticismo, geografia e collezionismo nel Rinascimento dei Paesi Bassi* (Modena: Franco Cosimo Panini, 1998).

8 Mangani, *Il mondo di Abramo Ortelio*; Denis Cosgrove, 'Globalism and tolerance in early modern geography', *Annals of the Association of American Geographers* 93 (2003): 852–70; Ann Blair, *The Theater of Nature: Jean Bodin and Renaissance Science* (Princeton: Princeton University Press, 1997).

9 Mangani, *Il mondo di Abramo Ortelio*, is a detailed study of these aspects of Ortelius's work.

10 Alessandro Scafi, *Mapping Paradise: A History of Heaven on Earth* (London: British Library, 2006). Giorgio Mangani, *Cartografia morale: geografia, persuasione, identità* (Modena: Franco Cosimo Panini, 2006), argues that moral considerations lie at the origins of all cartographic and geographic practice.

11 Matthew Edney, 'Reconsidering Enlightenment geography and map-making: reconnaissance, mapping, archive', in David N. Livingstone and Charles Withers (eds), *Geography and Enlightenment* (Chicago and London: University of Chicago Press, 1999), pp.165–98.

12 Luciana Martins, 'Mapping tropical waters', in Cosgrove, *Mappings*; Ronald E. Doel, Tanya J. Levin and Mason K. Marker, 'Extending modern cartography to the ocean depths: military patronage, Cold War priorities, and the Heezen Tharp Mapping Project, 1952–1959', *Journal of Historical Geography* 32 (2006): 605–26.

13 Bruno Latour, *Pandora's Hope: Essays on the Reality of Science Studies* (Cambridge, MA: Harvard University Press, 1999).

14 Hayden Lorimer and Katrin Lund, 'Performing facts: finding a way over Scotland's mountains', in Bronislaw Szerszynski, Wallace Heim and Claire Waterton (eds), *Nature Performed: Environment, Culture and Performance* (Oxford: Blackwell, 2003), pp.130–44; Sarah G. Cant, 'The tug of danger with the magnetism of mystery: descents into the comprehensive, poetic-sensuous appeal of caves', *Tourist Studies* 3 (2003): 67–81; John Wylie, 'Becoming icy: Scott and Amundsen's South Polar voyages', *Cultural Geographies* 9 (2002): 249–65; Neil Lewis, 'The climbing body nature and the experience of modernity', *Body and Society* 6 (2000): 58–80; John Wylie, 'An essay on ascending Glastonbury Tor', *Geoforum* (2002) 33: 441–54.

15 Mario Bagioli, *Galileo, Courtier: The Practice of Science in the Culture of Absolutism* (Chicago: University of Chicago Press, 1993); Eileen Reeves, *Painting the Heavens: Art and Science in the Age of Galileo* (Princeton: Princeton University Press, 1997).

16 Dava Sobel, *Longitude: The True Story of a Lone Genius Who Solved the Greatest Scientific Problem of His Time* (New York: Walker, 1995).

17 Martins, 'Mapping tropical waters'.

18 Anuradha Mathur and Philip da Cunha, *Deccan Traverses: The Making of Bangalore's Terrain* (Bangalore: Eastern Book Corporation, 2006).

19 Klaus Dodds, *Pink Ice: Britain and the South Atlantic Empire* (London and New York: I. B. Tauris, 2002). On the problems of field survey and sight in Antarctica see also William Fox, *Terra Antarctica: Looking into the Emptiest Continent* (San Antonio, TX: Trinity University Press, 2005).

20 Edmunds V. Bunkse, 'Saint-Exupéry's geography lesson: art and science in the creation and cultivation of landscape values', *Annals of the Association of American Geographers* 80 (1990): 96–108.

21 Antoine de St Exupéry, *Le Petit Prince* (Paris: L'Harmattan, 1943) (my translation).

22 Denis Guedj, *The Measure of the World: A Novel*, trans. Arthur Goldhammer (Chicago: University of Chicago Press, 2001), p.42.

23 Koch, *Cartographies of Disease*.

24 Ibid., pp.45–7. Malgaigne's map is reproduced on page 46.

25 David Blackbourn, *The Conquest of Nature: Water, Landscape, and the Making of Modern Germany* (New York: W.W. Norton, 2006).

26 Denis Cosgrove and Simon Rycroft, 'Mapping the modern nation', *History Workshop Journal* 40 (1995): 91–105.

27 Interview with Ian McHarg, available at <http://www.geoplace.com/gw/2001/0601/0601mem.asp>

28 John Cloud, 'American cartographic transformations during the Cold War', *Cartography and Geographic Information Science* 29 (2002): 261–82.

29 Denis Cosgrove, 'Contested global visions: One-World, Whole-Earth, and the Apollo space photographs', *Annals of the Association of American Geographers* 84 (1994): 270–94.

30 Latour, *Pandora's Hope*.

31 Frank Lestringant, *Mapping the Renaissance World: The Geographical Imagination in the Age of Discovery* (Berkeley and Los Angeles: University of California Press, 1994), pp.111–12.

Chapter 10: Carto-city

1 Adam T. Smith, *The Political Landscape: Constellations of Authority in Early Complex Polities* (Berkeley: University of California Press, 2003).

2 Most histories of urban form are illustrated with maps and plans but rarely explore critically the precise relations between these images and the built form of cities. See for example Richard Sennett, *Flesh and Stone: The Body and the City in Western Civilization* (New York: W. W. Norton, 1994); Joseph Rykwert, *Seduction of Place: The City in the Twenty-first Century* (New York: Pantheon, 2000).

3 Tom Koch, *Cartographies of Disease: Maps, Mapping and Medicine* (Redlands, CA: Esri Press, 2005), pp.105–24.

4 For historical examples, see Ola Söderström, 'Paper cities: visual thinking in urban planning', *Ecumene* 3 (1996): 249–81. For a more contemporary discussion of the processes of negotiating urban space planning through maps see Ola Söderström, Elena Cogato Lanza, Roderick J. Lawrence and Gilles Barby, *L'Usage du projet* (Lausanne: Editions Payot, 2000).

5 On Los Angeles planning ideas, see Greg Hise, *Magnetic Los Angeles: Planning the Twentieth Century Metropolis* (Baltimore: Johns Hopkins University Press, 1997) and William Deverell, *Eden by Design: The 1930s Olmsted–Bartholemnew Plan for the Greater Los Angeles Region* (Berkeley: University of California Press, 2000). On the use of ZIP codes, see Michael Curry and David Phillips, 'Privacy and the phenetic urge: geodemographics and the changing spatiality of local practice', in David Lyon (ed.), *Surveillance as Social Sorting: Privacy, Risk, and Automated Discrimination* (London: Routledge, 2003), pp.137–52.

6 The classic texts on these post-modern urban landscapes include Edward Relph, *The Modern Urban Landscape* (Baltimore: Johns Hopkins University Press, 1987); David Harvey, *The Condition of Postmodernity: An Enquiry into the Origins of Cultural Change* (Oxford and New York: Blackwell, 1989). Edward Soja, *Thirdspace: Journeys to Los Angeles and Other Real-and-Imagined Places* (Oxford: Blackwell, 1996) and *Postmetropolis: Critical Studies of Cities and Regions* (Oxford: Blackwell, 2000).

7 Despite the graphic sophistication of iconic communication in the contemporary city, the volume of text in the public landscape seems to increase with the number of enterprises competing for the attention of a highly mobile, car-borne public. This is very noticeable in the USA, where neon and other forms of illuminated text are characteristic features of urban space. On the role of text in regulating urban legibility in the past, see Rose Marie San Juan, *Rome: A City Out of Print* (Minneapolis: University of Minnesota Press, 2001). We should also recall the significant role traditionally given to public maps in countries such as Italy where they have frequently been carved onto the facades of public buildings.

8 Juergen Schulz, 'Jacopo de' Barbari's view of Venice: map making, city views and moralized geography before the year 1500', *Art Bulletin* 60 (1978): 425–74.

9 One of the principal advantages of the coordinate system for mapping is that it allows for greater mobility of the map than a purely pictorial outline and plot. Obviously, errors of transcription can occur in both forms, but the transmission of tables of figures from which a map may be constructed allows for much greater accuracy of positioning than merely copying lines and points by eye.

10 See the fascinating discussion of the grid as a mode of spatial representation in Edward Casey, *Representing Place: Landscape Painting and Maps* (Minneapolis: University of Minnesota Press, 2002), p.199–215.

11 Vaughan Hart and Peter Hicks, *Paper Palaces: The Rise of the Architectural Treatise* (New Haven: Yale University Press, 1998).

12 Manfredo Tafuri, *Architecture and Utopia: Design and Capitalist Development* (Cambridge, MA: MIT Press, 1976).

13 Lucia Nuti, 'Mapping places: chorography and vision in the Renaissance', in Denis Cosgrove (ed.), *Mappings* (London: Reaktion Books, 1999), pp.90–108;

Lucia Nuti, *Ritratti della città: visione e memoria tra medioevo e settecento* (Venezia: Marsilio, 1996).

14 Nicholas Crane, *Mercator: The Man Who Mapped the Planet* (London: Weidenfeld and Nicolson, 2002), 188.

15 For a detailed analysis of relations between urban mapping and costume books, see Barbara Wilson, *The World in Venice: Print, the City and Early Modern Identity* (Toronto: University of Toronto Press, 2005).

16 Kirsten A. Seaver, 'Norumbega and Harmonia Mundi in sixteenth-century cartography', *Imago Mundi* 50 (1998): 34–58.

17 Lorraine Daston and Katharine Park, *Wonders and the Order of Nature, 1150–1750* (New York: Zone Books, 1994).

18 G. Palsky, *Des Chiffres et des cartes. Naissance et développement de la cartographie quantitative au XIXe siècle* (Paris: CTHS, 1996).

19 The urban planner John Reps has devoted his career to the study of these urban images. See for example John W. Reps, *Lithographs of Towns and Cities in the United States and Canada, Notes on the Artists and Publishers, and a Union Catalog of Their Work, 1825–1925* (Columbia and St Louis: University of Missouri Press, 1984).

20 The history of the A–Z guide is told by its creator in Phyllis Pearsall, *A–Z Maps: The Personal Story from Bedsitter to Household Name* (London: Geographer's A–Z Map Company, 1990).

21 Denis Cosgrove and Simon Rycroft, 'Mapping the modern nation', *History Workshop Journal* 40 (1995): 91–105.

22 Tafuri, *Architecture and Utopia*.

23 Michael Silver, 'Mapping curves', in Michael Silver and Diana Balmori (eds), *Mapping in the Age of Digital Media: The Yale Symposium* (London: Wiley-Academy, 2003), pp.40–7.

24 Kevin Lynch, *The Image of the City* (Cambridge, MA: MIT University Press, 1960); Kenneth E. Boulding, *The Image: Knowledge in Life and Society* (Ann Arbor: University of Michigan Press, 1956).

25 Guy Debord, *The Society of the Spectacle* (Detroit: Black and Red, 1973). There is a very substantial literature on Situationism. Within geography a key text is David Pinder, 'Subverting cartography: the situationists and maps of the city', *Environment and Planning D: Society and Space* 28 (1996): 405–27. More recent writings on urban experience such as Steve Pile, *Real Cities: Modernity, Space and the Phantasmagorias Of City Life* (London: Sage, 2005), owe their inspiration to the Situationist focus on the relationship between urban experience and built form.

26 Michel de Certeau, *The Practice of Everyday Life* (Berkeley: University of California Press, 1985); Rosalyn Deutsche, 'Boy's town', *Environment and Planning D: Society and Space* 9 (1991): 5–30.

27 Michael Silver and Diana Balmori, 'Networking maps: GIS, virtualized reality and the World Wide Web', in Silver and Balmori, *Mapping in the Age of Digital Media*, pp.48–50.

28 The project may be viewed at <http://www.aroundgroundzero.net/>

Chapter 11: Seeing the Pacific

1 Denis Cosgrove and Veronica della Dora, 'Mapping global war: Los Angeles, the Pacific and Charles Owens' pictorial cartography', *Annals of the Association of American Geographers* 95/2 (2005): 373–90.

2 William L. Thomas, 'The Pacific basin: an introduction', in Herman Friis (ed.), *The Pacific Ocean: A History of Its Geographical Exploration* (New York: American Geographical Society, 1967), pp.1–17.

3 Bernard Smith, *European Vision and the South Pacific* (New Haven: Yale University Press, [1960] 1985). The historical geography of the oceans has become a focus of interest among geographers very recently and will undoubtedly continue to rework this conventional historiography. See the special issue of the *Journal of Historical Geography* 32/3 (2006) devoted to the oceans and edited by David Lambert, Luciana Martins and Miles Ogborn.

4 Matthew Fontaine Maury, *The Physical Geography of the Sea* (London: n.p., 1855); Graham Burnett, 'Matthew Fontaine Maury's 'Sea of Fire': hydrography, biogeography, and Providence in the Tropics', in Felix Driver and Luciana Martins (eds), *Tropical Views and Visions* (Chicago: University of Chicago Press, 2005), pp.113–36.

5 Matthew Fontaine Maury, *Maury's New Elements of Geography for Primary and Intermediate Classes* (New York: American Book Company, 1913); Hildegarde Hawthorne, *Matthew Fontaine Maury: Trail Maker of the Seas* (New York and Toronto: Longman, Green & Co., 1943).

6 Kenneth J. Bertrand, 'Geographical exploration by the United States', in Friis, *Pacific Ocean*, pp.292–320.

7 Quoted in Foster Rhea Dulles, *America in the Pacific: A Century of Expansion* (Boston and New York: Houghton Mifflin, 1932).

8 Frederick Jackson Turner, *Rise of the New West, 1819–1829* (New Haven: Yale University Press, 1906).

9 James C. Malin, 'Reflections on the closed space doctrines of Turner and Mackinder and the challenge of those ideas by the air age', *Agricultural History* 18 (1944): 65–74; 107–26.

10 Warren Zimmerman, *First Great Triumph: How Five Americans Made Their Country a World Power* (New York: Farrar, Straus and Giroux, 2002), pp.89–91.

11 Alfred Thayer Mahan, *The Influence of Sea Power upon History, 1660–1783* (Mineola, NY: Dover Publications, [1890] 1987).

12 Halford J. Mackinder, 'The geographical pivot of history' [1904], in *Democratic Ideals and Reality: With Additional Papers* (New York: W. W. Norton, 1962).

13 Ibid., pp.244, 248.

14 Quoted in Dulles, *America in the Pacific*, p.15.

15 Ibid., p.5.

16 Alfred T. Mahan, *The Interests of America in Sea Power* (Freeport, NY: Books for Libraries, [1897] 1970), p.243.

17 Alfred T. Mahan, *Naval Strategy* (Washington, DC: FMFRP 12–32, US Marine Corps, [1911] 1991), p.197.

18 Alfred T. Mahan, *Interests of America*, p.42. The collection includes 'Hawaii and our future sea power', written in 1893.

19 Mahan, *Interests of America*, p.31.

20 Rolf Hobson, *Imperialism at Sea: Naval Strategic Thought, the Ideology of Sea Power, and the Tirpitz Plan, 1875–1914* (Boston: Leiden, Brill, 2002), p.165.

21 The Chinese term *Yazhou*, or 'Asian continent', was itself a nineteenth-century neologism, a response to Western orientalist writing that replaced the long-standing imperial designation of China as *Zongguo* ('central kingdom') with *tianxia* ('all under heaven'), but that retained Chinese centrality by stressing continentality. Chinese spatiality had been accepted in Tokugawa Japan, placing the islands in a peripheral position on a Sino-centric globe. See Kevin M. Doak, 'Narrating China, ordering East Asia: the discourse of nation and ethnicity in imperial Japan', in Kai-wing Chow et al. (eds), *Construction of Nationhood in East Asia* (Ann Arbor: University of Michigan Press, 2001), pp.85–113.

22 Stefan Tanaka, *Japan's Orient: Rendering Pasts into History* (Berkeley and Los Angeles: University of California Press, 1993).

23 Susan Schulten, *The Geographical Imagination in America, 1880–1950* (Chicago and London: University of Chicago Press, 2001), p.17.

24 Ibid., p.33.

25 Ibid., p.34.

26 Ibid., pp.101–6.

27 Arnold Guyot, *Physical Geography* (New York and Chicago: Ivison, Blakeman, Taylor and Company, 1885).

28 Schulten, *Geographical Imagination*, p.71.

29 Guyot, *Physical Geography*, n.p.

30 'Francis Fontaine Maury', in Dumas Malone (ed.), *Dictionary of American Biography* (New York: Scribners, 1961), pp.428–31.

31 Matthew Fontaine Maury, *Physical Geography* (revised edn; New York: University Publishing Company, 1887).

32 Schulten, *Geographical Imagination*, p.141; Walter Gailey Hoffman, *Pacific Relations: The Races and Nations of the Pacific Area and Their Problems* (New York and London: McGraw Hill, 1936).

33 On flight and American popular culture see Robert Wohl, *The Spectacle of Flight: Aviation and the Western Imagination, 1920–1950* (New Haven: Yale University Press, 2005), esp. pp.9–48.

34 Hendrik Willem van Loon, *The Story of the Pacific* (New York: Harcourt, Brace and Company, 1940).

35 Ibid., p.384.

36 Denis Cosgrove, *Apollo's Eye: A Cartographic Genealogy of the Earth in the Western Imagination* (Baltimore: John Hopkins University Press, 2001), pp.243–8.

37 Archibald MacLeish, 'The image of victory', in Hans W. Weigert and Vilhjalmur Stefansson (eds), *Compass of the World: A Symposium on Political Geography* (London: George G. Harrap, 1946), pp.1–11.

38 Richard Edes Harrison, *'Look at the World': The Fortune Atlas for World Strategy* (New York: Alfred A. Knopf, 1944).

39 Ibid., p.41.

40 Ibid.

41 'Pacific battlefield: from the Moon it covers half the Planet', *Life*, 22 December (1941): 59–73. References on pp.62, 59.

42 Ibid., p.63.

43 *Los Angeles Times*, 17 April 1944.

44 Anne-Robert Jacques Turgot's idea that as one travels away from Europe one moves backwards through the stages of human social evolution was widely held well into the twentieth century and still informs popular stereotypes of 'primitive' peoples, including South Sea islanders. See Michael Heffernan, 'On geography and progress : Turgot's *plan d'un ouvrage sur la géographie politique* (1751) and the origins of modern progressive thought', *Political Geography* 13 (1994): 328–43.

Chapter 12: Seeing the Equator

1 The lines are taken from Edmund Spenser's *The Fairie Queene* (London: Dent, 1987), and form part of the epigraph to Herman Melville's *Encantadas* (see note 23 below).

2 Denis Cosgrove, *Apollo's Eye: A Cartographic Genealogy of the Earth in Western Imagination* (Baltimore: Johns Hopkins University Press, 2001), pp.29–53.

3 Isaac Newton, *Philosophiae Naturalis Principia Mathematica* (London: Jussu Societatis Regiae ca typis Josephi Streatii, 1687).

4 Originally Köppen took 5°C as the measure but this was later reduced to 3°, narrowing the width of the equatorial climatic belt. Variation in temperature at the Equator over the course of a year is caused by the earth's elliptical path around the sun, which brings the two bodies closest in early January (*perihelion*) and separates them by the greatest distance in early July (*aphelion*).

5 On climates and related living species, see Barry Lopez, *Arctic Dreams: Imagination and Desire in a Northern Landscape* (New York: Scribner, 1986), pp.15–41.

6 David N. Livingstone, 'The moral discourse of climate: historical consideration on race, place, and virtue', *Journal of Historical Geography* 17 (1991): 413–34; and 'Human acclimatization: perspectives on a contested field of inquiry in science, medicine and geography', *History of Science* 25 (1987): 359–94. It is possible that this myth arose from observation of the affects of the passage between hemispheres on animals that have a seasonal cycle of fertility. Ewes, for example, change their breeding season in response to the altered patterns of daylight.

7 Michael Taussig, 'The beach (a fantasy)', in W. J. T. Mitchell (ed.), *Landscape and Power* (Chicago and London: University of Chicago Press, [1994] 2002), pp.317–48.

8 The painting, once owned and celebrated by John Ruskin in *Modern Painters*, now hangs in the Boston Museum of Fine Arts.

9 Mark Twain, *More Tramps Abroad* (Leipzig: Tauchnitz, 1897), p.240.

10 Mark Twain, *Following the Equator: A Journey around the World* (London: Chatto and Windus, 1900).

11 Thurston Clarke, *Equator: A Journey round the World* (London: Hutchinson, 1988), p.17.

12 The terms general and specific geography are taken from the geographer Varenius. The former refers to geographical study of phenomena distributed across many different places (mountains, populations numbers, etc.), the latter to the study of unique places and regions.

13 Von Humboldt's map was of Mount Chimborazo, Ecuador's highest volcanic peak, which lies some 4° south of the Equator. Africa's highest peak, Mount Kilimanjaro, is a similar distance south of *The Line*.

14 See Michael Dettelbach, 'Global physics and aesthetic empire: Humboldt's physical portrait of the tropics', in David Miller and Peter Reill (eds), *Visions of Empire: Voyages, Botany, and Representations of Nature* (Cambridge: Cambridge University Press, 1996).

15 Gerry Kearns, 'The imperial subject: geography and travel in the work of Mary Kingsley and Halford Mackinder', *Transactions of the Institute of British Geographers* 22 (1997): 450–72.

16 'Anthropophagi' is the name used on maps and in descriptive texts of the tropics up to the eighteenth century, as I discuss in Chapter 3.

17 Felix Driver and Luciana Martins (eds), *Tropical Views and Visions* (Chicago: University of Chicago Press, 2005); David Arnold, 'Inventing tropicality', in *The Problem of Nature: Environment, Culture and European Expansion* (Oxford: Blackwell, 1996), pp.141–68; Roy MacLeod and Milton Lewis (eds), *Disease, Medicine and Empire: Perspectives on Western Medicine and the Experience of European Expansion* (London: Routledge, 1988).

18 One might mention also the Rwanda genocide in the early 1990s and the civil wars of Liberia and Sierra Leone, which, despite the equivalent barbarity of contemporaneous 'ethnic cleansing' in the Balkans, only confirmed long-standing visions of peculiar violence at the Equator.

19 Jared Diamond, *Collapse: How Societies Choose or Fail to Survive* (London: Allen Lane, 2005).

20 John R. Gillis, *Islands of the Mind: How the Human Imagination Created the Atlantic World* (New York: Palgrave Macmillan, 2004). See also the special issue on islands in *Journal of Historical Geography* 29 (1993): 487–660.

21 Marco Squarcini, 'Equator', opening essay in Gian Paolo Barbieri, *Exotic Nudes* (Köln/London: Taschen, 2003), n.p.

22 France continued to conduct nuclear tests in the tropical Pacific into the 1990s.

23 Herman Melville, *The Encantadas, or Enchanted Isles* (New York: Harper and Brothers, 1962[1854]), 157–225; quotation on p.160.

24 Ibid., p.161.

25 Ibid., p.167.

26 Ibid., p.163.

27 Ibid., p.178.

28 Ibid.

29 Sarah Radcliffe, 'Imaginative geographies, post-colonialism and national identities: contemporary discourses of the nation in Ecuador', *Ecumene* 3 (1996): 21–42.

30 Clarke, *Equator*, pp.254–5.

Index